Data-Enabled Approaches for Enhancing the Air Force Transformational Capability Pipeline

MATTHEW WALSH, LANCE MENTHE, JONATHAN L. BROSMER, JULIE GEORGE, CHRISTINE KISTLER LACOSTE, ERIC HASTINGS, SHERRILL LINGEL, ÉDER SOUSA

Prepared for the Department of the Air Force
Approved for public release; distribution unlimited

PROJECT AIR FORCE

For more information on this publication, visit **www.rand.org/t/RRA952-1**.

About RAND

The RAND Corporation is a research organization that develops solutions to public policy challenges to help make communities throughout the world safer and more secure, healthier and more prosperous. RAND is nonprofit, nonpartisan, and committed to the public interest. To learn more about RAND, visit www.rand.org.

Research Integrity

Our mission to help improve policy and decisionmaking through research and analysis is enabled through our core values of quality and objectivity and our unwavering commitment to the highest level of integrity and ethical behavior. To help ensure our research and analysis are rigorous, objective, and nonpartisan, we subject our research publications to a robust and exacting quality-assurance process; avoid both the appearance and reality of financial and other conflicts of interest through staff training, project screening, and a policy of mandatory disclosure; and pursue transparency in our research engagements through our commitment to the open publication of our research findings and recommendations, disclosure of the source of funding of published research, and policies to ensure intellectual independence. For more information, visit www.rand.org/about/research-integrity.

RAND's publications do not necessarily reflect the opinions of its research clients and sponsors.

About This Report

The Air Force 2030 Science and Technology (S&T) Strategy calls for enhanced investment in transformational technologies to enable new warfighting concepts with leap-ahead capabilities (Department of the Air Force [DAF], 2019). In response, the DAF has dedicated a portion of the S&T budget to the transformational component of the portfolio and has created the Transformational Capabilities Office (TCO) within the Air Force Research Laboratory (AFRL) to manage these activities. The mission of the TCO is to foster transformational capabilities across a range of programs. A key question in this regard is how to identify and evaluate the most promising concepts, and how to do so in a data-enhanced manner.

To this end, the TCO asked RAND Project AIR FORCE (PAF) to develop data-enhanced approaches to identify, select, and promote the most promising concepts to feed into the transformational capability pipeline. In this report, we propose a human-centered, data-enhanced (HCDE) decisionmaking process. The process is intended to answer the question of which technologies to develop to meet the needs of the current and future force. The process is especially relevant to the TCO given the TCO's mandate to broadly consider capability gaps and technology solutions, yet elements of the process can be adopted by organizations with different responsibilities. The process leverages a prototype data science tool, the Semantic Clustering Analysis and Thematic Exploration Tool, for extracting information from free-text data sources. The process also leverages a collection of future-focused techniques, called foresight methods, that allow human experts to combine information from data sources, along with domain knowledge and creativity, to arrive at potential solutions. We demonstrate the HCDE process in three case studies involving (1) high-speed vertical takeoff and landing, (2) Joint All-Domain Command and Control, and (3) human capital management.

The research reported here was commissioned by the AFRL's TCO (AFRL/RS) and conducted within the Force Modernization and Employment Program of PAF as part of a fiscal year 2021 add-on project, "Data-Driven Approaches for Advancing the Air Force Transformational Capability Pipeline."

RAND Project AIR FORCE

PAF, a division of the RAND Corporation, is the DAF's federally funded research and development center for studies and analyses, supporting both the U.S. Air Force and the U.S. Space Force. PAF provides the DAF with independent analyses of policy alternatives affecting the development, employment, combat readiness, and support of current and future air, space, and cyber forces. Research is conducted in four programs: Strategy and Doctrine; Force

Modernization and Employment; Resource Management; and Workforce, Development, and Health. The research reported here was prepared under contract FA7014-16-D-1000.

Additional information about PAF is available on our website: www.rand.org/paf/

This report documents work originally shared with the DAF on December 1, 2021. The draft report, issued on October 11, 2021, was reviewed by formal peer reviewers and DAF subject matter experts.

Acknowledgments

We would like to thank Christopher Ristich (the director of the AFRL TCO), and our action officers, David Myers and Kathy Bauman (AFRL/RS) for their assistance and guidance.

We also thank the many military personnel, civilians, and stakeholders whose views informed the research conducted for this project, and especially those in the AFRL Aerospace Systems Directorate; the AFRL Airman Systems Directorate/711th Human Performance Wing; the AFRL Commander's Accelerated Initiatives Office; the AFRL Information Directorate; the AFRL Materials and Manufacturing Directorate; the AFRL Munitions Directorate; the AFRL Plans and Programs Directorate; the AFRL Space Vehicles Directorate; the AFRL Technology Directorate AFWERX; the AFRL TCO leads for the WARTECH and Explore programs; Air Combat Command: Futures; Air Combat Command: Science and Technology; the Air Force Directorate of Studies, Analyses, and Assessments; the Air Force Life Cycle Management Center Architecture and Integration Directorate; the Air Force Tactical Exploitation of National Capabilities program; the Defense Advanced Research Projects Agency Strategic Technology Office; the U.S. Department of Defense Joint Artificial Intelligence Center, the MITRE Corporation; the National Air and Space Intelligence Center Geospatial and Signatures Intelligence Group; the Pacific Air Forces: Science and Technology; the Pacific Air Forces Strategic Plans Division; RAND; RAND Australia; U.S. Army Research Development and Engineering Command; and the U.S. Army Research Laboratory.

We thank our reviewers, Chad Heitzenrater and Raymond Kuo, for their constructive feedback on earlier versions of this report. Finally, we thank the many other RAND colleagues who helped us with this work. Principally, but not exclusively, we thank Elizabeth Bodine-Baron, Jim Chow, Natalie Crawford, Frank Delsing (of the 2021 RAND Air Force Fellows Program), Miriam Matthews, Libby May, and Christopher Mazerluft.

Summary

Issues

- A key goal for the Transformational Capabilities Office (TCO) is to foster transformational capabilities across a range of initiatives.
- To propose, develop, and select which concepts to advance into the transformational capability pipeline, the TCO must extract information from data contained across many text sources that report capability gaps and technology programs, and combine it with human expertise and creativity.
- Machine learning (ML) and natural language processing (NLP) can be used to extract information from text sources on capability gaps and technology solutions.
- Subject matter expertise must also be captured and leveraged effectively to provide creative insight and make best use of the extracted information.

Approach

To understand how human-centered, data-enhanced (HCDE) decision processes can be used to determine which concepts to advance into the transformational capability pipeline, we used a multimethod qualitative approach that included a review of the relevant bodies of literature on development planning, along with interviews with senior leaders, technical experts, and subject matter experts from the Department of the Air Force (DAF) and the U.S. Department of Defense (DoD). The synthesis of our analysis revealed opportunities for the TCO to (1) use data science tools to extract information from vast databases of capability gaps, capability needs, and technology solutions; and (2) use a more diverse set of future-focused decision methods, called foresight methods, to leverage human expertise and creativity. We developed and implemented a proof-of-concept software tool, the Semantic Clustering Analysis and Thematic Exploration Tool, to extract information from free-text descriptions of capability gaps and technologies, and we combined the tool with foresight methods as part of an HCDE decisionmaking process. We demonstrate the data science tool and foresight methods in three case studies involving (1) high-speed vertical takeoff and landing, (2) Joint All-Domain Command and Control, and (3) human capital management.

Key Findings

- The TCO has an exceptionally broad mandate that includes performing open-ended searches across operational capability gaps and technology solutions. This calls for tools and methods different from those used by other DAF and DoD organizations.
- Some data sources for capability gaps are widely referenced, but they are not centrally managed. Data sources for science and technology (S&T) solutions are far more varied and diverse. The volume of data contained across these sources is vast.

- No software tools are systematically used to parse, extract, and summarize the content of capability gap and technology solution sources.
- Modern data science techniques can be used to extract information from free-text descriptions contained in these sources.
- Development planning is a human-centered endeavor that depends on domain knowledge, creativity, and social networks.
- Foresight methods can be used to leverage human expertise and creativity.
- Data science techniques and foresight methods can be integrated to form an HCDE decisionmaking process that involves
  - using data science methods to extract an initial set of capability gaps from data sources that contain formal guidance and/or operational experiences
  - using human-centered methods to iterate the selection of capability gaps and to enrich their descriptions
  - using data science methods to extract an initial set of technology solutions from S&T sources
  - reiterating with human-centered methods to determine the final technology solution set to be proposed.

Recommendations

1. The AFRL and the TCO should use the concept development and selection process described above.
2. The AFRL and the TCO should use a software tool, like the one described in this report, to extract information from natural language data sources. As they do so, they should conduct user testing and validation studies to improve the software tool.
3. The AFRL should explore alternate NLP methods to maximize the utility of information extracted from free-text sources.
4. The DAF should curate and standardize key operational capability gap data sources.
5. The AFRL, the DAF, and the TCO should enrich key S&T data sources by purchasing or developing capabilities to cleanse records and merge them with metadata.
6. The TCO should expand the use of creative, interactive, expert-driven, and evidence-based foresight methods.
7. As a stepping stone to reach full curation and standardization of HCDE capability development planning, the AFRL and the TCO should record human-generated technology pairings for capability gaps.

Contents

Figures and Tables

Figures

Tables

Chapter 1. Introduction

Multiple developments in the global security environment are prompting the Department of the Air Force (DAF) and the U.S. Department of Defense (DoD) to take action to guard the technological superiority of U.S. military forces. First, as the DAF and DoD transition from supporting military operations against nonstate actors to preparing for great-power competition, capability needs are changing: Technologies developed to support lengthy counterinsurgency/counterterrorism operations in a permissive environment will not be sufficient to support rapid major combat operations in a denied environment. Second, great-power competitors are aggressively modernizing their air, space, and cyber capabilities. Third, many of today's science and technology (S&T) advances originate from outside the DoD and are available to competitors through the global marketplace.

In response to this new strategic environment, the DAF is changing how it manages S&T. One of the primary objectives for the Air Force S&T enterprise, as laid out by the former Secretary of the Air Force, Heather Wilson, in the Air Force Science and Technology 2030 Strategy, is to "develop and deliver transformational strategic capabilities." This objective calls for the DAF to allocate a dedicated portion of the S&T budget (20 percent) to the transformational component of the portfolio. The transformational component is focused on system-of-systems technologies to enable new warfighting concepts (DAF, 2019, p. iii).

In the fall of 2019, the Air Force technology executive officer (TEO) stood up the Transformational Capabilities Office (TCO) within the Air Force Research Laboratory (AFRL) to manage the transformational component of the DAF S&T portfolio. Table 1.1 describes some of the major initiatives that the TCO manages. Both the Seedlings for Disruptive Capabilities and the Explore initiatives establish pathfinders to discover, assess, and mature transformational capabilities. The WARTECH initiative provides a forum for operational subject matter experts (SMEs) and technologists to propose, validate, and refine capability requirements for future force design. Finally, the Vanguard initiative provides enterprise support to develop and transition a limited number of high-priority transformational capabilities to the DAF.[1]

---

[1] The Vanguard initiative reflects an enterprise-wide commitment to delivering capabilities that transform operations for future forces. The initial set of Vanguard programs, as outlined in the S&T 2030 strategy document (DAF, 2019), included Skyborg (for unmanned, low-cost wingmen), Golden Horde (for networked, precision-guided weapons), and NTS-3 (for improved position, navigation, and timing using software defined radios); see AFRL, undated. Although the Vanguard programs represent strategic investments, they constitute only about 20 percent of the core DAF S&T budget allocated to the transformational component, with the remainder going to other 6.2 (applied research) and 6.3 (advanced technology development) efforts within the AFRL and 6.3 efforts elsewhere throughout the DAF (U.S. Air Force Scientific Advisory Board, 2020).

Table 1.1. Major Transformational Capabilities Office Initiatives

Initiatives	Description	Outcomes
Seedlings for Disruptive Capabilities	AFRL-initiated, cross-disciplinary applied research programs to advance innovative science for transformational capabilities	Maturation of innovative science to transformational capabilities
Explore	Commercial market requests accessible to entire national technology ecosystem to develop multidisciplinary solutions for future force capability challenges	DAF business case for market transformational capabilities
WARTECH	Continuous annual process to codevelop proposals with S&T and operational communities for transformational capabilities to address capability requirements derived from future force design concepts and DAF strategic guidance	Validated and mature concepts for future force transformational capability needs
Vanguard	AFWIC/AQR/PEO/TEO/user partnerships to develop and deliver game-changing capabilities to transform DAF operations	Proof-of-concept and transition preparation for transformational capabilities

SOURCE: U.S. Air Force Scientific Advisory Board, 2020.
NOTE: AFWIC = Air Force Warfighter Integration Capability; AQR = Air Force Acquisition: Science, technology, and Engineering Directorate; PEO = program executive officer.

To support the objective of developing and delivering transformational strategic capabilities to the DAF, the TCO asked RAND Project AIR FORCE to recommend data-enhanced approaches to identify, select, and promote the most promising concepts to feed a technology pipeline that enables new warfighting concepts with leap-ahead—or transformational—capabilities.

The resulting proposal for a human-centered, data-enhanced (HCDE) decision process is highly relevant to the TCO given the TCO's mandate to broadly consider capability gaps (i.e., statements of current or future operational or technology needs) and technology solutions. By using natural language processing (NLP) and machine learning (ML) to extract information from free-text data, the HCDE decision process broadens the aperture of capability gaps and technology solutions that the TCO may consider. The process is also responsive to the need to identify transformational technologies. By using future-focused foresight methods, HCDE leverages the knowledge and creativity of human experts to arrive at innovative and transformational capabilities. Although the process is especially relevant to the TCO, other organizations responsible for identifying or prioritizing capability gaps and technology solutions may nonetheless benefit from using elements of the process.

Study Context

This study is positioned at the convergence of themes related to development planning and transformational capabilities.

Development Planning

To maintain air, space, and cyber superiority, the DAF must continuously update its technology in response to evolving threats and emerging opportunities. In some cases, incremental adjustments are enough; in other cases, transformational change is needed. The latter requires clear and specific strategic guidance, senior-leader support, high risk tolerance, research and development (R&D) investments, and timely and adaptable acquisition processes. *Development planning* is the set of analytic activities that the DAF uses to anticipate, prioritize, and pursue R&D opportunities and to acquire materiel capabilities to support the National Defense Strategy (NDS).[2]

The DAF has recently taken steps to invigorate development planning. The *USAF Strategic Master Plan* of 2015 emphasized its importance, stating, "the capability development process itself must also become more responsive, adaptable, and agile" (DAF, 2015). To create a more responsive capability development process, the U.S. Air Force (USAF) established Air Force Futures to conceptualize and shape future force design along with the Strategic Development Planning and Experimentation office to provide analytic support for development planning (Under Secretary of the Air Force and Vice Chief of Staff, 2017). Most recently, the TEO stood up the TCO as a sister organization to the office to oversee Vanguards—flagship efforts to develop and deliver leap-ahead capabilities—along with other initiatives that contribute to transformational strategic capabilities (see Table 1.1). The TCO works closely with Air Force Futures and AQR to develop concepts for future force design and to acquire transformational capabilities.

Transformational Capabilities

In the past, the DAF and DoD S&T enterprises have identified conceptual classes of technologies with revolutionary potential.[3] Disruptive or game-changing technologies are technologies that introduce new system attributes or capabilities rather than simply improving upon existing ones (Christensen, Raynor, and McDonald, 2015; DAF, 2015; National Research Council, 2010). This is sometimes referred to as a nonlinear capability gain. Offset strategies are strategies centered on technologies that change the nature of competition rather than simply increasing strength in an existing area.[4] Disruptive or game-changing technologies have this strategic potential.

---

[2] For a detailed review of AF development planning, see Knopman et al., 2020; Leftwich et al., 2019; and National Research Council, 2014.

[3] Revolution in military affairs may be brought about by changes in technology, doctrine, strategy, tactics, or force structure (Metz and Kievit, 1995). The AFRL seeks to bring about revolution primarily through technology.

[4] Previous and current offset strategies have emphasized nuclear weapons, stealth technology, precision-guided weapons, and artificial intelligence (AI).

The S&T 2030 strategy document (DAF, 2019) introduced the new concept of *transformational capabilities*, which are distinguished by the following features:

- They are future focused (i.e., future force design through leap-ahead experimentation and prototyping).
- They are cross-disciplinary (i.e., multidisciplinary problem solving and system-of-systems thinking).
- They are warfighter driven (i.e., capabilities directly informed by operational warfighter needs).
- They are timely (i.e., technology advances rapidly transformed into operational concepts).[5]

Figure 1.1 shows how transformational capabilities relate to game-changing/disruptive capabilities and enduring technologies—that is, technologies related to foundational DAF S&T missions. Transformational capabilities are distinct in that they lay at the intersection of cross-discipline solutions to provide nonlinear capability gains in the short term. Game-changing/disruptive technologies and enduring technologies may have some, but not all, of these attributes. For example, game-changing/disruptive technologies may also be delivered in the long term and they may leverage single-discipline solutions, and enduring technologies produce linear capability gains.

Figure 1.1. Relationships Between Enduring, Game-Changing/Disruptive, and Transformational Technologies

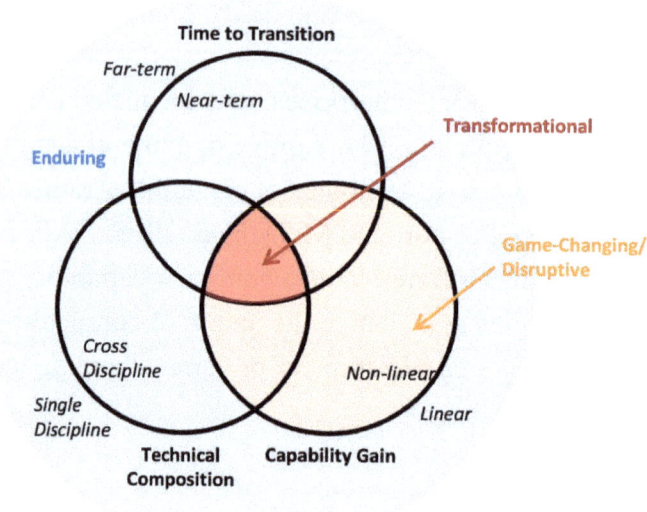

---

[5] While development planning spans 20 years, transformational capabilities mainly lie within the first five-year increment.

A Human-Centered, Data-Enhanced Decision Process to Inform Transformational Concepts

Figure 1.2 shows a high-level HCDE decision process for selecting the most promising concepts to feed a transformational technology pipeline. The figure depicts three general sets of activities: (1) identifying capability gaps; (2) identifying S&T programs; and (3) prioritizing which potential S&T solutions to pursue.[6] A capability gap may lead to the initiation or selection of S&T programs that could feasibly address the need. In the commercial sector, this is called "market pull" (Brem and Voigt, 2009). Alternatively, S&T programs may be initiated without a clear linkage to an existing capability gap. These technologies may enable future-focused capabilities not yet conceived of. In the commercial sector, this is referred to as "technology push." Hence, the relationship between the first and second sets of activities—identifying capability gaps and identifying S&T programs—is bidirectional. In a resource-constrained environment, decisionmakers must end by prioritizing potential S&T solutions and deciding which to advance into the transformational capability pipeline.

Figure 1.2. The Human-Centered, Data-Enhanced Decision Process

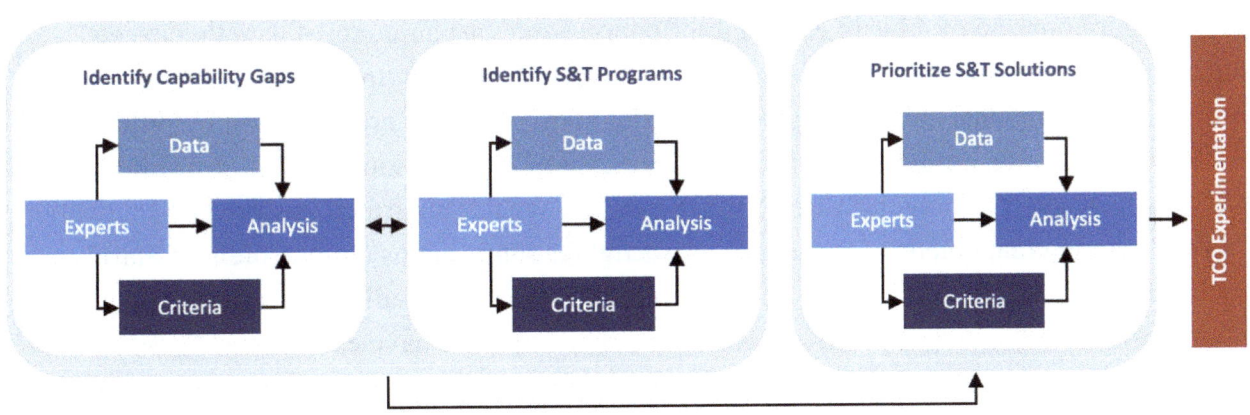

A common set of elements underlie each set of activities: Human experts generate and identify relevant data sources along with criteria for evaluating capability gaps, S&T programs, and S&T solutions. They then apply a range of analytic techniques to assess the criteria and to select which capability gaps, S&T programs, and S&T solutions to pursue.

---

[6] Within the defense acquisition community, the term *capability gap* sometimes refers to an operational need (e.g., the ability to perform a certain mission), and it sometimes refers to a technical need (e.g., the ability for a weapon system to meet certain performance parameters). The term *program* sometimes refers to the major unit of aggregation used for accounting (e.g., a program element), and it sometimes refers to the subordinate unit of aggregation used to describe a related set of S&T activities. The meaning of these terms throughout this report depends on the usage.

AI is a discipline concerned with machines demonstrating intelligence—that is, behaving in seemingly rational ways given what they know (Russel and Norvig, 1995). ML is a subfield of AI that involves training machines to perform tasks without first providing explicit instruction (Jordan and Mitchell, 2015). NLP is also a subfield of AI, one that involves giving computers the ability to understand and generate language (Cambria and White, 2014). Some, but not all, NLP approaches use ML (Young et al., 2018).

Algorithmic breakthroughs in these areas, along with increases in computing power and data availability, have contributed to a surge of interest in AI. This is evident in the commercial sector. For example, about half of executives who responded to a recent McKinsey global survey reported that their companies had incorporated AI into at least one core business function (Balakrishnan et al., 2020). This is consistent with findings from the Stanford Institute for Human-Centered Artificial Intelligence's *Artificial Intelligence Index Report 2022* (Zhang et al., 2022). This trend is also evident within the DoD. For example, the fiscal year (FY) 2019 National Defense Authorization Act established the Joint Artificial Intelligence Center to coordinate the DoD's efforts to develop AI technologies for warfighters, as well as the National Security Commission on Artificial Intelligence to ensure global leadership in AI (Public Law 115-232, secs. 238, 1051).

One common use of AI is to extract information from vast amounts of data that exceed human processing capacity.[7] This is relevant to development planning because operational limitations, operational capability gaps, and technology programs are described in millions of records distributed across dozens of sources. For example, crowdsourced platforms like the Joint Lessons Learned Information System (JLLIS) and AFWERX's Airmen Powered by Innovation (API) program contain tens of thousands of warfighter submissions (Joint Chiefs of Staff, undated). Additionally, as the DAF seeks to expand work with small businesses and other nontraditional defense contractors, more than 1,300 companies have entered the AFWERX portfolio, and the number and diversity of Small Business Innovation Research (SBIR) and Small Business Technology Transfer (STTR) awards has grown tremendously (Lofgren, 2020). Lastly, as the pace of science has accelerated, it has become more difficult to track advances that may enable new technologies. The arXiv repository for scientific papers contains nearly 2,000,000 articles and receives nearly 20,000 new submissions each month, and thousands of National Science Foundation (NSF) awards are granted annually (Cornell University, 2021).

All these sources primarily contain free-text data. By using NLP to extract information from the data, along with ML to organize the information, the DAF can leverage these sources to enhance development planning.

---

[7] Examples are analyzing terabytes of data from wide-area surveillance sensor systems or analyzing millions of consumer transactions.

The Role of Human Expertise and Creativity in an HCDE Process

Innovation, or the implementation of a new or significantly improved product, process, or service, depends on human expertise and creativity (Kline and Rosenberg, 2010). Innovation requires, among other things, an understanding of the technology outputs of S&T activities, an understanding of end-user needs, products or processes that leverage S&T outputs in new ways to meet end-user needs, and the marketing of products and processes to increase end-user adoption.

R&D is a critical for technology innovation. However, the rapid growth and globalization of S&T activities complicates R&D management in two ways (Cozzens et al., 2010). First, the rate of knowledge generation is so high that it has become difficult to track even relatively narrow research areas; this is a *processing challenge*. Second, the rate of discovery is so high that it has become difficult to anticipate future opportunities and disruption; this is a *prediction challenge*.

A variety of methods may be applied to R&D management. AI and ML can address the data processing challenge. Foresight methods—techniques for using critical thinking, planning, and management competencies to evaluate the impact of long-term uncertainties on short-term decisions—can address the prediction challenge (Popper, 2008). The federal government and the DoD already use many types of foresight methods (Greenblott et al., 2019). Common examples include conferences and workshops, scenario planning, modeling and simulation, and trend analysis. In fact, backcasting, a foresight method that involves defining a future state and working backward through the sequence of advances and changes needed to get there from the present, was central to development planning during the 1990s (Knopman et al., 2020).

R&D management and innovation are still predominantly human-driven processes. Human experts must establish a future vision, identify data sources and metrics, interpret outputs of data analyses, make decisions about R&D priorities and actions, and ultimately link technology innovations with end users. Notwithstanding the human-driven nature of these processes, data science methods may expand the breadth, depth, and speed of R&D management and innovation.

Integration of Human and Machine Intelligence in an HCDE Process

Figure 1.2 shows the integration of human and machine intelligence in an HCDE decision process. Given the size of data sources, analytic methods—including AI, ML, and NLP—may be needed to extract information from the sources. Human experts then conduct analyses to evaluate capability gaps, S&T programs, and S&T solutions according to the predefined criteria. These analyses may take the form of quantitative methods, such as modeling and simulation, or they may take the form of other qualitative or semiquantitative methods—for example, conferences and workshops or scenarios.

Here is one specific example of how this process may play out:

1. SMEs from Major Commands (MAJCOMs) and other operational communities identify capability gaps. They generate multiple data sources including formal integrated priority lists (IPLs) and informal warfighter descriptions of capability limitations and lessons learned. The capability gaps are then evaluated according to multiple criteria including

operational significance. These evaluations are performed using wargaming, along with other methods. Presently, only one or a limited number of capability gaps are considered. The collection of capability gaps is too large and distributed for human SMEs to consider in its entirety. Data science methods could potentially allow SMEs to detect patterns and extract information from a larger set of capability gaps.

2. SMEs from academia, industry, and DAF S&T organizations identify S&T programs. Many of these programs are described in sources like the Defense Technical Information Center (DTIC), arXiv, and program budget documents. Associated technologies are evaluated according to multiple criteria, including technology readiness level (TRL). These evaluations may use horizon scanning and landscape analysis along with other techniques. Presently, only a limited number of S&T programs are considered. Once again, the collection of S&T programs is too large and distributed for human SMEs to consider in its entirety. Data science methods could potentially allow SMEs to identify other promising technology solutions.

3. SMEs from Headquarters, Air Force A5 (Strategy, Integration, and Requirements), Headquarters, Air Force A9 (Studies, Analyses, and Assessments) and other offices prioritize S&T solutions. Data may be available from modeling and simulation. Solutions are evaluated according to multiple criteria, including operational advantage. These evaluations may involve additional modeling and simulation, quantitative scenarios, and other methods.

The goal of this study was to operationalize the collection of activities shown in Figure 1.2, including recommending HCDE science methods to realize an HCDE process.

Study Methodology

To understand how the TCO can combine data sources with human expertise in an HCDE decision process, we used a multimethod approach, as shown in Figure 1.3.

Figure 1.3. Study Approach

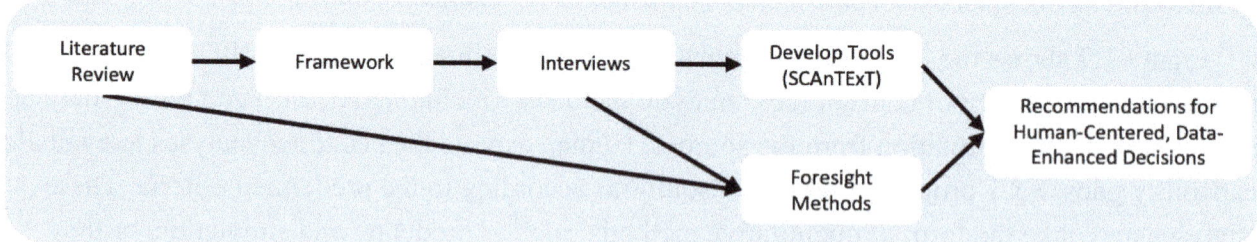

We first identified how the DAF and DoD prioritize R&D activities today. We conducted thirty semistructured discussions with senior leaders, technical experts, and SMEs from the AFRL, the DAF, and the DoD. These interviews traced the major sets of activities shown in Figure 1.2.

Next we explored the feasibility and utility of using data science approaches to extract information from sources that contain free-text descriptions of operational capability gaps, operational needs, and technology solutions. We implemented a prototype tool, the Semantic Clustering Analysis and Thematic Exploration Tool (SCAnTExT) to allow an analyst to extract information from these sources. We applied the tool to a collection of more than 20 data sources that comprised hundreds of thousands of records.

We then reviewed literature on foresight methods. Based on an analysis of the specific goals and constraints present for different TCO initiatives, we recommend "playbooks" of different foresight activities for each program. Many of the methods make use of information extracted from the data sources.

Finally, we applied SCAnTExT, along with different foresight methods, to three case studies involving (1) high-speed vertical takeoff and landing (HSVTOL); (2) Joint All-Domain Command and Control (JADC2); and (3) AI for human resource management. From these activities we drew recommendations for an HCDE decision process.

The Organization of This Report

The remainder of this report is organized as follows:

- Chapter 2 presents results from interviews with senior leaders, technical experts, and SMEs about capability development planning in the Air Force and the DoD.
- Chapter 3 describes a detailed HCDE process that emerged from the interviews.
- Chapter 4 describes data sources that contain textual descriptions of operational capability gaps and technology solutions, an NLP path that prepares those data sources for ML, and SCAnTExT, the software application that allows an analyst to interact with those data sources.
- Chapter 5 reviews work on foresight methods and recommends playbooks of methods for different TCO initiatives based on their goals and constraints.
- Chapter 6 demonstrates the use of SCAnTExT and other foresight methods in the three case studies.
- Chapter 7 summarizes the work and provides recommendations.

Chapter 2. Interviews

To understand how the DAF and DoD currently identify capability gaps, identify potential technological solutions, and prioritize development efforts, we interviewed more than 30 senior leaders and SMEs from over 25 defense S&T organizations.[1] Besides establishing a baseline, the purpose of these interviews was to uncover how elements of HCDE decisionmaking can be incorporated into development planning. Organizations in interviews include:

- AFRL Aerospace Systems Directorate
- AFRL Airman Systems Directorate/711th Human Performance Wing
- AFRL Commander's Accelerated Initiatives Office
- AFRL Information Directorate, and its deputy capability lead for Air Combat Command (ACC)
- AFRL Materials and Manufacturing Directorate
- AFRL Munitions Directorate
- AFRL Plans and Programs Directorate
- AFRL Space Vehicles Directorate
- AFRL Technology Directorate AFWERX: spark cells, strategic innovation, Agility Prime program
- AFRL Transformational Capabilities Office: WARTECH and Explore programs
- Air Combat Command: Futures (Advanced Battle Management System/JADC2)
- Air Combat Command: Science and Technology
- Air Force Life Cycle Management Center Architecture and Integration Directorate
- Air Force Studies, Analysis, and Assessments
- Air Force Tactical Exploitation of National Capabilities
- Defense Advanced Research Projects Agency Strategic Technology Office
- DoD Joint Artificial Intelligence Center
- MITRE Corporation: data science program
- Intelligence Group
- National Air and Space Intelligence Center Geospatial and Signatures
- Pacific Air Forces Strategic Plans Division
- RAND Australia: critical technologies of national interest
- U.S. Army Research, Development, and Engineering Command
- U.S. Army Technology Forecasting Office

In these discussions we looked to understand current practices—typical methods, unusual methods, what works, and what does not—and to identify commonly used data sources and software tools. We also sought to capture any other lessons learned that might inform how the

---

[1] We conducted more than 30 interviews, but some interviewees belonged to the same organizations.

TCO should go about proposing, developing, and prioritizing concepts to advance into the transformational capability pipeline.

To identify consistent findings from these interviews, we used *thematic analysis* (Attride-Stirling, 2001). This is a method commonly used to identify, organize, and report patterns (i.e., themes) found within a large body of data in the social sciences (Braun and Clarke, 2006). We coded relevant statements—those that appeared likely to answer our research questions—systematically across 225 transcribed comments.

Table 2.1 lists the resulting 13 themes grouped into four major categories. *Note: To encourage frank discussions, all interviews were conducted with the promise of complete confidentiality, so no quotations or findings will be attributed to any specific organizations or interviewees.*

Table 2.1. Overview of Interview Responses

Major Category		Theme
Organizational scope	1.	Some organizations seek S&T solutions for a constrained set of operational capability gaps; others seek operational capability gaps that can be addressed by a constrained set of S&T solutions.
	2.	This creates a divide between organizations that identify capability gaps and ones that identify technology solutions.
	3.	Social networks are a primary means to deal with the divide.
Identification of capability gaps	4.	Organizations rely on MAJCOMs and Combatant Commands (CCMDs) to provide "top-down" guidance in IPLs and capability gap lists (CGLs).
	5.	Organizations also rely on interpersonal communication among operators, analysts, and SMEs to augment "top-down" guidance with "bottom-up" input.
	6.	Formal capability gaps must first be translated to make them S&T actionable.
	7.	No software tools are systematically used to parse, extract, or summarize operational capability gap data sources.
Identification of technology solutions	8.	Most organizations use SMEs to identify promising technologies.
	9.	Searches are, however, human intensive, potentially incomplete, and prone to bias.
	10.	DAF engagement with industry has increased the flow of ideas into the AFRL.
	11.	Nascent tools for mining S&T data sets exist but are not yet widely used.
Technology transfer	12.	Organizations recognize the importance of involving operational and acquisition communities throughout development.
	13.	The challenge of bridging the "valley of death" remains.

Organizational Scope

Theme 1. Most Organizations Perform a Constrained Search

Based on our review of the literature, we established a three-stage model for capability development and decisionmaking (see Figure 1.2: *identify capability gaps, identify potential technological solutions*, and *prioritize development efforts*). However, when asked to describe the relative weights of their organization's efforts across these three sets of activities, almost all defense S&T organizations described being constrained in their roles such that their decision process is effectively reduced to a two-stage model, as shown in Figures 2.1 and 2.2.

Figure 2.1. The Constrained Decision Process: Problem-Focused

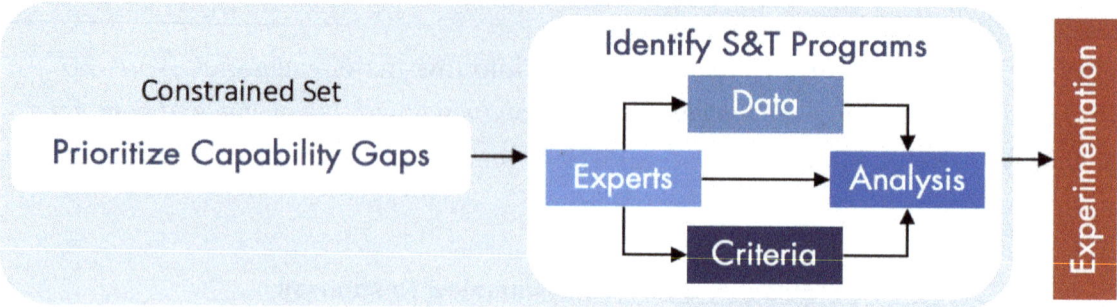

Figure 2.2. The Constrained Decision Process: Solution-Focused

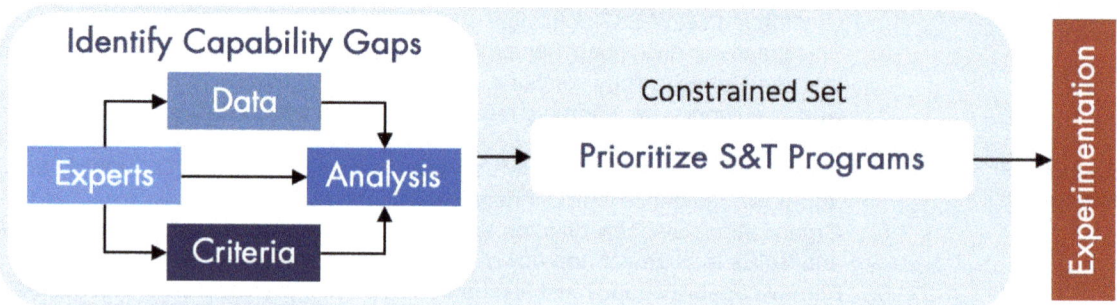

In other words, some defense S&T organizations are designed to focus on a **specific problem set**—for example, improving communications or sensors. As one interviewee stated, "We're not shy about saying to a vendor, 'Great technology, but it has no current applicability here.'" Other defense S&T organizations are designed to focus on a **specific solution set**—for example, AI or directed energy. As another interviewee remarked, "We had a capability that was being developed. . . . I didn't have a requirement or capability gap, but everyone recognized the military utility." As a result, most of these organizations implicitly approach capability development and decisionmaking in one of two ways: either searching for a promising technology that might fill a given capability gap or searching for a critical capability gap that a given technology might fill.

This observation is in no way meant to disparage these organizations. On the contrary, their respective constraints reflect a clear understanding of enduring defense needs and promising technology areas. The point is that the TCO is different. The TCO has an unusually broad scope in that it is charged to look comprehensively at both the problem space and the solution space.

As the AFRL company commander, Maj Gen Heather Pringle, described it to the Senate Armed Services Subcommittee on Emerging Threats and Capabilities, the TCO is designed to "lead the transformational S&T portfolio and work across technology directorates to develop multidisciplinary, system-of-systems solutions" (Pringle, 2021, p. 4).

Thus, while the TCO can learn—and benefit—from the experiences of both problem- and solution-focused organizations, to accomplish its mission the TCO also needs unique tools, techniques, and procedures that are not found in most defense S&T organizations.

Theme 2. There Is a Divide Between Operational-Focused and S&T-Focused Organizations

The insular focus of defense S&T organizations creates a divide between identifying capability gaps and identifying S&T programs that may address them. On the one hand, organizations focused on operations may not grasp the capabilities and limitations of certain technologies. As one interviewee observed, "I think AFRL has to fill in the blanks for [Air Force] Futures—provide S&T knowledge." On the other hand, organizations focused on S&T may not grasp operational needs and employment. As another interviewee noted, "The lab develops weapons to enhance warfighters . . . but a lot of researchers without an operational background struggle with the tactics." One consequence of the divide is that operators may have unmoored expectations of some technologies, and they may overlook other technologies entirely. Another consequence is that technologists may overestimate the ease of transition or utility for some use cases, and they may overlook other use cases entirely.

Theme 3. Social Networks Are a Primary Means to Deal with the Divide

Many S&T organizations stated that social networks are essential for working across the divide. For example, "It's a talk, it's a monthly call with ACC, PACAF and EUCOM [European Command], it's relationships developed over the years. I want to spend time keeping up with them." Or, "The way cross-organizational contacts happen is largely through personal relationships." It is encouraging that individuals in S&T organizations invest so greatly in social networks; this was the most common theme from the interviews. At the same time, several interviewees remarked about high turnover rates among staff. This creates a liability if social networks are the primary means for dealing with the divide between identifying capability gaps and identifying S&T programs that may address them.

To some extent, individuals and organizations will always have relatively greater depth of knowledge in one of these two areas. Yet the AFRL and the DAF are undertaking steps to bridge the divide. For example, the AFRL allows researchers from technical directorates to temporarily work in the TCO and with Air Force Futures, and a key feature of TCO initiatives (e.g., Vanguards) is to allow technologists and operators to work together.

Identification of Capability Gaps

Theme 4. Organizations Rely on Major Commands and Combatant Commands for "Top-Down" Guidance

We found significant consistency in the data sources and methods used by defense S&T organizations to identify capability gaps. These organizations typically extract capability gaps from formal statements of needs from MAJCOMs and CCMDs for "top-down" guidance. As one interviewee explained, "We look at IPLs, STIPLS [science and technology integrated priority lists], and formal requirements statements. Those are nuanced. They are narrowly focused, and by and large, those statements of capability gaps are formal." Another explained, "Identifying capability gaps is not easy, but it's straightforward. We have a lot of them [capability gaps]."

Some DAF organizations also draw on Air Force Futures products for guidance, although these do not yet appear to have achieved the widespread currency of IPLs. Organizations frequently mentioned the 2018 NDS, but there was consensus that the NDS offered little utility in terms of identifying specific technical gaps. As one interviewee put it bluntly, "[T]he NDS is not actionable [from an S&T perspective]."

Theme 5. Organizations Also Rely on Communications with Operators for "Bottom-Up" Inputs

While most organizations reported drawing on IPLs and related strategic guidance, they also draw on interpersonal communication among operators, analysts, and SMEs to augment these with "bottom-up" inputs. One interviewee explained, "The key to make all this easier is to maintain those close relationships—proving your value and role to the warfighter." Another summed it up this way: "Identifying warfighter needs happens through close engagement with the warfighter and the senior military folks at the MAJCOMs and CCMDs. It's top-down *and* bottom-up." Another emphasized the importance of looking outside the staff in general: "You're not going to get the same answer from the operators that you're going to get from someone like me that's been sitting on the staff for 20 years." These data sources are seen as being complementary. Speaking of the relationship between formal top-down guidance and informal bottom-up inputs, one interviewee remarked, "[T]hey're often very aligned. At the warfighter S&T level, it's about putting specificity to it."

Theme 6. Formal Capability Gaps Must Be Refined to Make Them S&T Actionable

Notably, while S&T organizations draw on formal statements of need from MAJCOMs and CCMDs, they must first refine these statements to make them S&T actionable. As one interviewee noted, "The CGL as written is going to immediately point to kinetic or directed energy." In other words, a CGL or IPL may identify a capability without considering other means to achieve the desired effect, or the accompanying technologies needed to deliver that capability. Moreover, one interviewee noted that "the gaps are not one-technology specific . . . and it's getting worse with complexity of systems." Indeed, the mapping from capability gap to technology solution is one to many. S&T organizations must convert CGLs and IPLs into STIPLs to apply technical solutions. As one interviewee described it, their task is to "do that enrichment . . . interpret what they (operators) really want or need and break it down for SMEs in the technical directorates."

Theme 7. No Software Tools Are Systematically Used to Store and Analyze Capability Gaps

Although we found strong commonality in data sources and approaches, we found no evidence of software tools to parse, extract, or summarize these data sources in any automated fashion. This is particularly problematic for crowdsourced databases that contain many warfighter submissions. Referring to one such database, an interviewee remarked, "We have this system that end users are putting data in. Getting the data has been the LIMFAC [limiting factor]." In Chapter 4 we describe a data tool that we developed for this purpose. Based on our interviews, we believe that this tool may be of value not just to the TCO but also to other solution-focused defense S&T organizations, such as the AFRL and the U.S. Army Research Laboratory.

Identification of Technology Solutions

Theme 8. Most Organizations Use SMEs to Identify Promising Technologies

Unlike the challenge of identifying capability gaps, which is generally retrospective in nature, the challenge of identifying new technologies is prospective or predictive. The output is therefore less certain and the guidance from above less authoritative. Perhaps for this reason, we found no common data sources used by defense S&T organizations for this purpose. We did, however, find strong commonality in terms of process: most S&T organizations solicit qualitative input from technologists and other SMEs.

To identify technology solutions, most organizations use a handful of familiar methods: working groups, whiteboard sessions, tabletop exercises, and—most commonly—informal conversations with representatives from industry and academia, often as "targets of opportunity" at events or conferences. One interviewee explained the process this way: "We periodically convene whiteboard sessions as a *coalition of the willing*; we invite specific people that may have particular insights and bring in colored pens and stickies and unpack a problem or technology." Another explained it in broader terms: "Engagement with industry—that's where you understand new ideas and challenges to making the technology a reality. You can leverage a lot from the industry base. Bringing in diversity of thought from different schools . . . to refresh your staff." A few organizations reported using surveys of SMEs, or bringing in guest speakers, but rarely did they report using more quantitative methods, such as the Delphi method.[2]

---

[2] The Delphi method is an interactive process used to arrive at consensus among a group of experts.

There was widespread recognition across organizations that it would be impossible to track all potentially relevant technical developments. One interviewee noted, "As an office, I have a couple of consultants on my staff to beat the bushes"; another said, "It's a bandwidth issue. I would love an office to run technologies through the grinder and dissect them and match them to gaps." To overcome this human capital limitation, some organizations mentioned using federally funded research and development centers, University Affiliated Research Centers, and consultants to perform landscape and horizon analyses.

Theme 9. Technology Searches Are Human Intensive and Potentially Incomplete and Biased

Many interviewees expressed their concerns about the inconsistencies in the processes they use. One expressed it this way: "We know a little bit about a lot. It's a lot of connecting the dots and matchmaking and talking to people." More than one interviewee noted that the outcome varied greatly depending on the specific SMEs involved. As one put it, "The right technologists are important because they have to identify the technology landscape"; a second agreed that it was "[a] strong process when you have the right people [to] support the process," and a third noted, "It's personality driven yet today, and not well documented."

Some also questioned the value of these exercises, raising concerns of confirmation bias and a potential echo chamber. As one interviewee put it candidly, the outcomes are "based on people's brilliance and the degree to which I agree with them. So, it's biased." Finally, others noted that when workshops are time intensive, they are a burden on SMEs, which can limit their participation and value. In general, there was a sense that the technology search process was at best difficult and incomplete.

As we describe in Chapter 3, a host of other foresight methods—which are not in use today—can be used to solicit more structured feedback from SMEs, and some of them may be well suited to TCO initiatives.

Theme 10. DAF Engagement with Industry Has Increased the Flow of Ideas into the AFRL

The DAF has sought to increase engagement with the small businesses ecosystem through the SBIR and STTR programs and, in recent years, with AFWERX. For example, AFWERX is in a unique position to "bring in nontraditional companies into the government and streamline acquisition process," according to one interviewee.[3] Through the SBIR and STTR programs, AFWERX is trying to break down barriers for smaller innovative companies to partner with the DAF and DoD at large. This has resulted in large numbers of ideas and proposals flowing from industry—and the volume of data is approaching the limit of human processing capacity. Several interviewees stated this good news/bad news story. For example, one said, "I get 100 companies

---

[3] AFWERX is a program office at the AFRL that connects innovators across government, industry, and academia.

a month that want to talk with me." Another noted, "What I'm experiencing in my role is that we are getting hit from multiple avenues . . . for those of us who are trying to survey the space and maintain awareness, it becomes difficult to manage." The DAF has succeeded at soliciting ideas from industry, but the challenge of extracting potential technology solutions from these inputs persists.

Theme 11. Nascent Tools for Analyzing S&T Data Exist but Are Not Yet Widely Used

We found instances of general-purpose tools used to exchange program data within organizations. As one interviewee noted, "In the past, we have run off spreadsheets that you share with your friends and colleagues." PowerPoint and SharePoint were also mentioned. We also found some instances of tools and approaches used for mining data sources for future technologies. Bibliometrics and patent analysis were identified as methods sometimes used to identify emerging technologies; bibliometrics recognizes trends in technological development by applying text analytics to a document (e.g., title and abstract).

These tools and approaches were infrequently used, however, perhaps out of concern that they do not add new information but simply reinforce what has already been learned through SME input. For example, one interviewee observed that "knowing what I know about NLP and data mining and the maturity of those tools . . . often only reinforces the innovation or insight we've already made, or it's at a level that's useful for the executive but not the SME." Another interviewee echoed this sentiment: "The tools are not that good. . . . It's like confirmation bias."

Technology Transfer

The final two themes that emerged from this series of interviews were unexpected and unsolicited. While the format of the semistructured interviews anticipated that the scope of the discussion would begin and end with the organizations' involvement in the capability development process, many interviewees drew our attention to issues parallel to, and subsequent to, this process. Some felt that these concerns were at least as significant as the ones we were attempting to address in this report. These related issues may be summarized as (1) *involving the operational and acquisition community* throughout the development process, and (2) *bridging the "valley of death"* for technologies between conception and maturation.

Theme 12. The Importance of Involving the Operational and Acquisition Communities

Most interviewees commented on the importance of involving the operational community throughout the R&D process to ensure that the technologies produced are properly integrated into the DAF at the end. One interviewee said plainly, "AFRL is better than industry, but they have no clue how to integrate. . . . They need to understand warfighter needs from the start." Tabletop exercises offer an opportunity for S&T teams to involve warfighters, some noted, but "getting qualified warfighters to technology games is the number one difficulty."

One benefit of involving the operational community is to discover weaknesses in concepts of operations (CONOPs)—an important element of developmental planning in the 1980s as well. Weaknesses may relate to the failure to anticipate an adversary's response. As one interviewee remarked, "We need to think more about red threats and *red-team the tech.*" Weaknesses may also relate to the failure to accurately represent the operational environment. Another interviewee stated, "We ask, what are the assumptions? And sometimes we just notice, that's not how the world works." In both cases, red teaming can reveal weaknesses in CONOPs still in the conceptual phase.

If the operational and S&T communities are two legs of the stool, the acquisition community is the third leg. Accordingly, many interviewees also noted the importance of involving the acquisition community as well. One interviewee observed, "I wish we would learn from that experience more broadly, so that every project has operations and intelligence, and acquisition and logistics." Speaking of the perceived success of the Air Reserve Weapons and Tactics Conference, another interviewee offered, "All of the players are in the same place; money guys, weapons guys, leadership to sign off."[4]

Theme 13. The Challenge of Bridging the "Valley of Death"

In capability development, the transition from S&T to advanced development has been characterized as a "valley of death,"[5] in which promising technologies fail to transition from basic research to an acquisition program (see, for example, Warwick, 2021). This occurs between DoD research, development, test, and evaluation (RDT&E) budget activity codes 6.3 (advanced technology development) and 6.4 (advanced component development).[6] Many interviewees expressed frustration with the DoD acquisition process. For example, one said, "[The] bottom line, our current acquisition process is broken and can't keep up with technology development."[7] Another said, "If you map the money, you can see the valley of death. . . . I don't know why we don't fix these obvious problems that we've known about for decades." Still other interviewees explained that "[t]he big gap is between TRL 4 and 7" and that "it's not an innovation problem, it's a transition problem." The reasons offered range from the difficulty of

---

[4] An acquisition plan is not the only transition pathway. Customer funding, part-of-transfer, and external collaborations may also lead to technology transition. These alternate pathways exist but did not come up in interviews.

[5] The language derives from a similar technology transfer problem in the civilian world between academia and industry: "In technology transfer, the 'valley of death' is the metaphor often used to describe the gap between academic-based innovations and their commercial application in the marketplace"; Gulbrandsen, 2009, p. 2.

[6] DoD RDT&E funding is characterized by the type of work performed. Budget activity codes 6.1–6.3 are referred to as the S&T budget. In contrast, budget activity codes 6.4, 6.5, and 6.7 refer to the application of existing science and technical knowledge to meet current or short-term operational needs. See, for example, Sargent, 2020, p. 3.

[7] For AFRL technology directorates, crossing this divide requires an acquisition plan. One interviewee explained, "We are trying to help [technology transfer], and it's through the planning and budgetary process . . . having that process, however detailed . . . you would need to find money for that [technology]. That's part of our process."

coordinating the pace of technology development with the Program Objective Memorandum (POM) cycle, shifting MAJCOM and CCMD priorities, and how RDT&E funds are allocated to achieve higher TRLs. Yet, regardless of the reason, recognition of the "valley of death" was among the most common themes expressed in the interviews.

Chapter Summary

Semistructured interviews with over 30 individuals from more than 25 S&T organizations provided insight into current practices for identifying and prioritizing operational capability gaps and technology solutions. Based on the 13 themes from our semistructured interviews, we identified both challenges and opportunities for HCDE processes for identifying transformational capabilities (see Table 2.1).

We found strong commonality in terms of data sources and methods used to identify capability gaps, but only weak commonality in terms of data sources and methods used to identify technology solutions. We found clear opportunities for data-enhanced decisionmaking tools and new foresight methods to wrangle SME input. We also found that significant technology transfer issues persist, and that if these are not resolved, the most effective TCO initiatives may ultimately go to waste.

Developing capabilities for future forces is and will continue to be a complex and time-consuming effort. There are uncertainties about future threat environments, emerging technologies and their development, budget constraints, and competing interests among different communities—R&D, acquisition, and warfighter—within the DAF. The DAF has had periods of success and failure in attempting to navigate these complexities in a manner that delivers capabilities at the needed time. In subsequent chapters we will describe creative foresight methods and new data tools that can help the TCO improve the capabilities development process.

Chapter 3. A Human-Centered, Data-Enhanced Approach to Identify and Prioritize Technology Concepts

Interviews with senior leaders and SMEs from diverse S&T organizations revealed three challenges that an HCDE decision process can help to address: (1) the divide that exists between organizations that are focused on capability gaps and technology solutions; (2) the human-intensive, incomplete, and potentially biased compilation of capability gaps and technology solutions; and (3) the human expertise and creativity needed to make capability gaps S&T actionable and to discover innovative solutions. In this way, an HCDE decision process can directly address themes related to organizational scope, identification of capability gaps, and identification of technology solutions.[1]

Figure 3.1 illustrates an HCDE process to address these limitations. The flow diagram represents the evolution of the more general framework shown in Figure 1.2, and it serves as a counterpoint to the more fractionated situation shown in Figures 2.1 and 2.2.

Figure 3.1. The Capability Development Decisionmaking Process

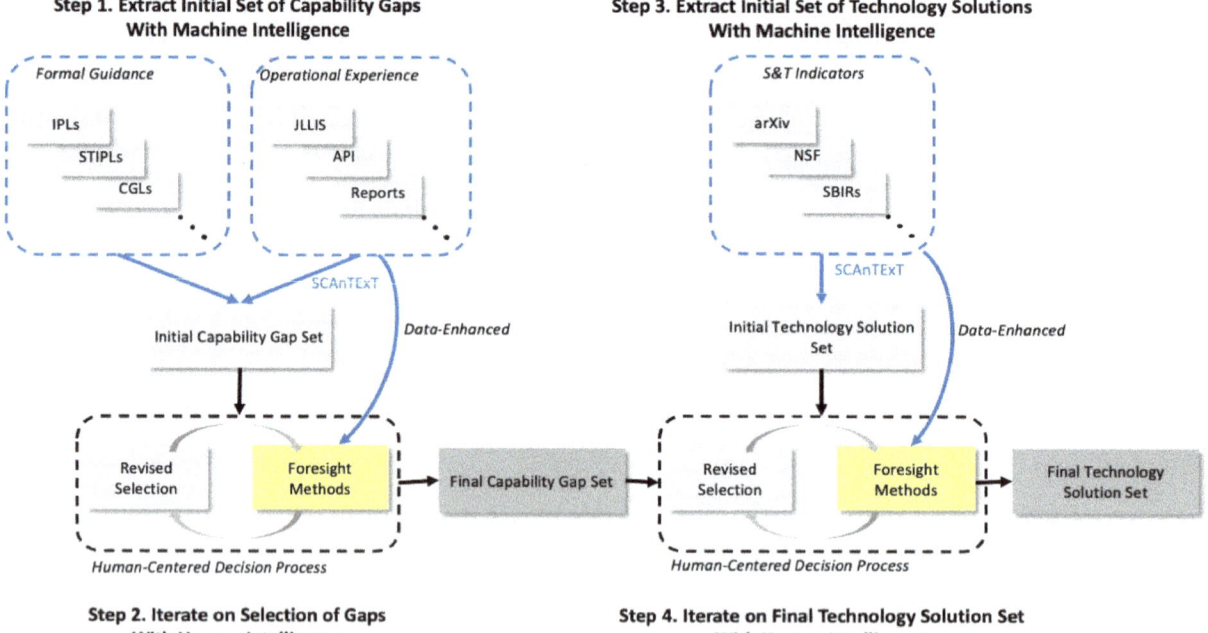

---

[1] An HCDE decision process may indirectly address themes related to technology transition, though additional steps are needed to address those challenges.

The process has four major steps:

1. using data science methods (i.e., ML and NLP) to extract an initial set of capability gaps from data sources that contain formal guidance and/or operational experiences
2. using human-centered methods (i.e., foresight methods) to iterate the selection of capability gaps and to enrich their descriptions
3. using data science methods to extract an initial set of technology solutions from various S&T sources, depending on the need
4. reiterate with (different) human-centered methods to determine the final technology solution set to be proposed.

The HCDE decision process takes advantage of the chief strength of the machine: to comprehensively search a vast database of capability gaps and technology solutions and to extract information from these sources. It also takes advantage of the chief strengths of the human: to assess the relevance of information returned by the machine, and to combine it into integrative technology solutions to operational capability gaps. In the chapters that follow, we advance from the abstract diagram in Figure 3.1 to a workable model with data science tools to extract information from textual sources and structured methods to leverage human expertise and creativity.

Application of an HCDE Process to the Transformational Capabilities Office

The major components in Figure 3.1 are conceptually aligned with the TCO WARTECH process, which begins from future force design threats and capability needs, advances to S&T proposal development, and concludes with prioritization of concepts. A diverse mix of SMEs, data sources, and analytic techniques are needed to identify the most promising concepts to feed into a transformational capability pipeline. The TCO is separate from the technical directorate structure of the AFRL and works closely with Air Force Futures. From this vantage point, the TCO can draw from diverse operational and S&T expertise.

The TCO must perform a relatively unconstrained search through many capability gaps and S&T programs originating in DoD, industry, and academia. These capability gaps and S&T programs are described using natural language. AI, ML, and NLP methods have shown potential to extract information from such sources. Additionally, the TCO must perform future-oriented technology analysis. Foresight methods are needed to account for uncertainty in future technology development and context when making short-term decisions.

The criteria used to evaluate capability gaps, S&T programs, and S&T solutions may be tailored for transformational capabilities. Significant issues and promising concepts that fail to meet these criteria may be directed elsewhere. For example, sources describing future force capability gaps (i.e., Air Force Futures) are most relevant to the TCO. While descriptions of current and short-term capability gaps (e.g., JLLIS and MAJCOM IPLs) are useful for establishing connections to warfighter needs, they may ultimately be directed to a contractor or program office. Likewise, technology solutions that are cross-discipline and approaching maturation are most relevant to the TCO. While S&T concepts that pertain to a single discipline or that have low TRLs may mature into transformational capabilities, they may be directed to the Air Force Office of Scientific Research, a technical directorate, or Strategic Development Planning and Experimentation in the short term.

The HCDE process is intended to help human decisionmakers answer the question of which technologies to develop to meet the needs of the current and future force. The process is highly relevant to the TCO given its mandate to broadly consider capability gaps and technology solutions. The process could be used, for example, to gather relevant inputs and to identify operational and S&T SMEs to support existing TCO programs. It could also be used to conceive of new programs by identifying significant capability gaps or emerging technology solutions. Finally, data-enabled elements of the HCDE process could be used to direct technology solutions to offices within the AFRL and DAF ecosystem with a need for those capabilities.

The process outlined in Figure 3.1 is not specific to the TCO or to transformational capabilities. Other organizations responsible for identifying or prioritizing capability gaps and technology solutions can borrow elements of the process while tailoring the data sources and criteria used to evaluate concepts.

Chapter Summary

An HCDE process can help to bridge the divides between capability gaps, technology solutions, and fielded systems. Such a process depends on the distinct affordances of human and machine intelligence. In the chapters that follow, we demonstrate a candidate HCDE process, and we underscore its utility.

Chapter 4. Data Science Methods to Support an HCDE Decision Process

A key element of an HCDE process is to use machine intelligence to extract information from vast data sources (i.e., a data-enabled decision process). In this chapter we consider the suitability of data science methods and data sources to support this dimension of HCDE decision processes.

To identify which transformational concepts to pursue, the TCO must track a vast array of operational capability gaps and technology solutions. Numerous data sources could enhance the selection of promising concepts. For example, CCMDs periodically publish IPLs, and additional capability gaps can be inferred from decentralized sources like warfighter submissions to the JLLIS. In addition, technology programs are described in sources like SBIR awards, NSF awards, the RDT&E Defense Budget, and journal publications. The amount of text data that the TCO would need to analyze to access this information presents challenges. Nonetheless, recent advances in NLP and the increase in cloud/distributed computing power could allow the TCO to leverage these data. NLP goes beyond simple word searches: by characterizing semantic content from text data, NLP methods allow records with similar themes and concepts to be grouped together and retrieved even if they do not share the same words.

NLP is a branch of AI that involves giving computers the ability to understand text and spoken language (Cambria and White, 2014). NLP combines computational linguistics, ML, and other statistical techniques to extract meaning from language. Common NLP tasks include speech recognition, part-of-speech tagging, name entity recognition, sentiment analysis, and natural language generation. These NLP capabilities underlie many commercial products and business processes. For example, NLP has been used to predict disease based on electronic health records, provide information about customer opinion, filter emails, and enable natural language interfaces like Amazon's Alexa and Apple's Siri (Young et al., 2018). In addition, ongoing advances in NLP, like the development of generative pretrained transformers, continue to enable new types of language-centered applications and business functions.[1]

In this chapter we explore the feasibility of using NLP to assist with the task of tracking operational capability gaps and technology solutions. The NLP path from data to solutions is shown in Figure 4.1. We begin by describing some relevant data sources for this work. We then discuss processing steps for data exaction, preparation, and semantic analysis. Finally, we present a software application that allows a user to interact with the NLP path to gather and synthesize information about capability gaps and technology solutions.

---

[1] Generative pretrained transformers use deep learning to learn about the contextual relationships between words in a text; see Wilson and Daugherty, 2020.

Figure 4.1. The Natural Language Processing Path

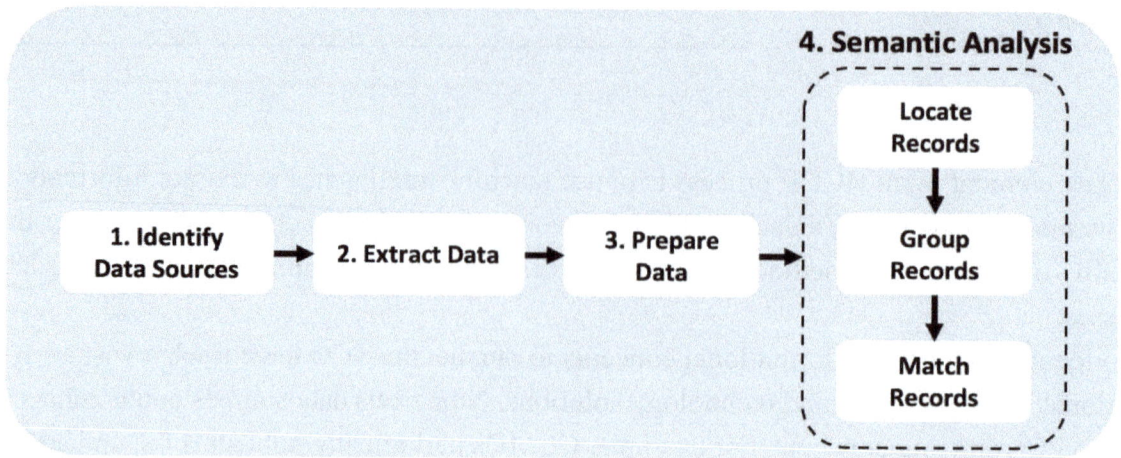

The Natural Language Processing Path

Data Sources

Operational Capability Gaps

The DAF does not maintain a central repository of operational capability gaps. Documents that describe operational needs are distributed—and take different forms—across commands, organizations, and offices. Given this, we began by searching for sources to compile a repository of operational capability gaps. These range from formal gaps identified in CCMD IPLs to informal gaps submitted to crowdsourced systems like API and JLLIS. Based on our interviews, members of the S&T community routinely use IPLs to understand capability gaps.[2] Although organizations infrequently reported using crowdsourced systems like API and JLLIS, they did note that some warfighter submissions to such systems provide context that can be helpful in terms of understanding the nuances of the capability gaps contained in CCMD IPLs.

Figure 4.2 shows how the capability gap data sources that we considered relate to one another. Formal capability gaps are derived from national military strategy, DAF objectives, and Joint Force objectives. Informal capability gaps are submitted by individuals in operational units. Although the two types of sources are not typically considered together, informal gaps may validate and enrich specific CCMD and MAJCOM capability gaps. Alternatively, many related informal gaps may signal the need for an as-of-yet unspecified formal gap.

---

[2] Interviewees also mentioned requirement mechanisms for specific communities, like Cyber Need Forms for cyber and the ISR Capabilities Analysis Requirements Tool for intelligence. Including these and additional sources would augment coverage of the capability gaps space.

Figure 4.2. Relationships Between Capability Gap Sources

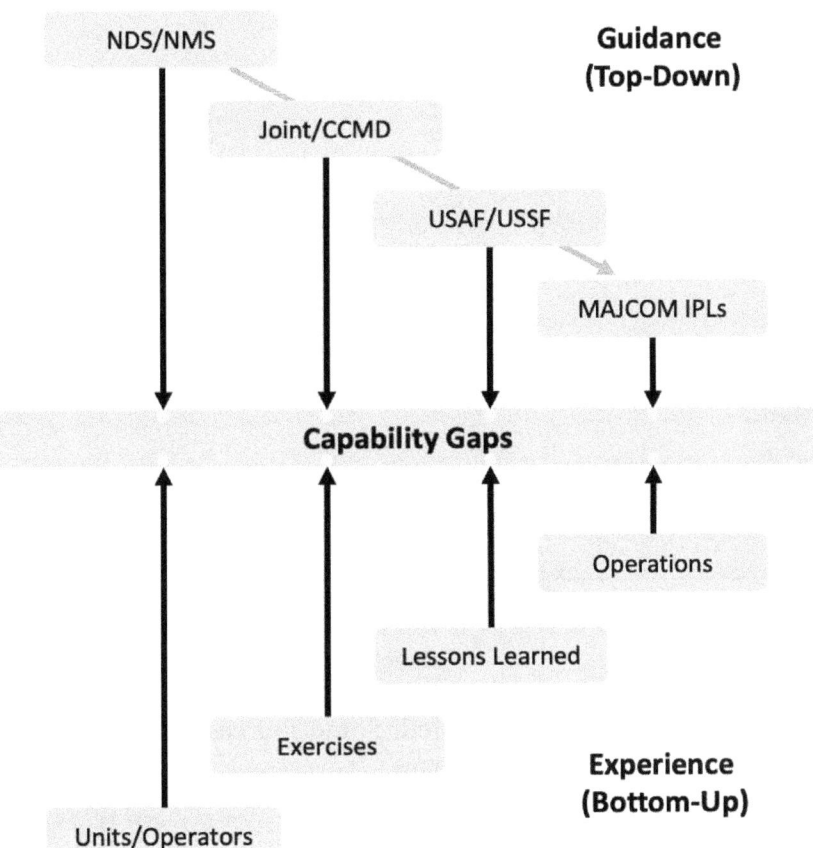

NOTE: NMS = National Military Strategy.

Technology Solutions

Potential technology solutions for operational capability gaps exist across a range of basic research activities, applied research activities, and system development and sustainment activities. Additionally, these technologies originate from government, industry, and academic sectors. Figure 4.3 shows how the technology data sources that we consider relate to one another along these dimensions. For example, arXiv is an open-access archive with approximately 2,000,000 scholarly articles that describe basic scientific research in academia. The NSF publishes abstracts for tens of thousands of awards that describe a mix of basic and applied research activities primarily conducted in academia. The U.S. government's SBIR website (SBIR.gov) publishes abstracts for tens of thousands of SBIR and STTR awards that describe applied research and system development activities in industry and academia. AFVentures curates a database with over 1,000 company biographies. Finally, the RDT&E Defense Budget, along with associated POM submissions, describes DoD-wide investments in basic research, applied research, and system development and sustainment activities. Each of these data sources contains different information and can potentially give a different view of the maturity and capabilities of a given technology.

Figure 4.3. Relationships Between Science and Technology Sources

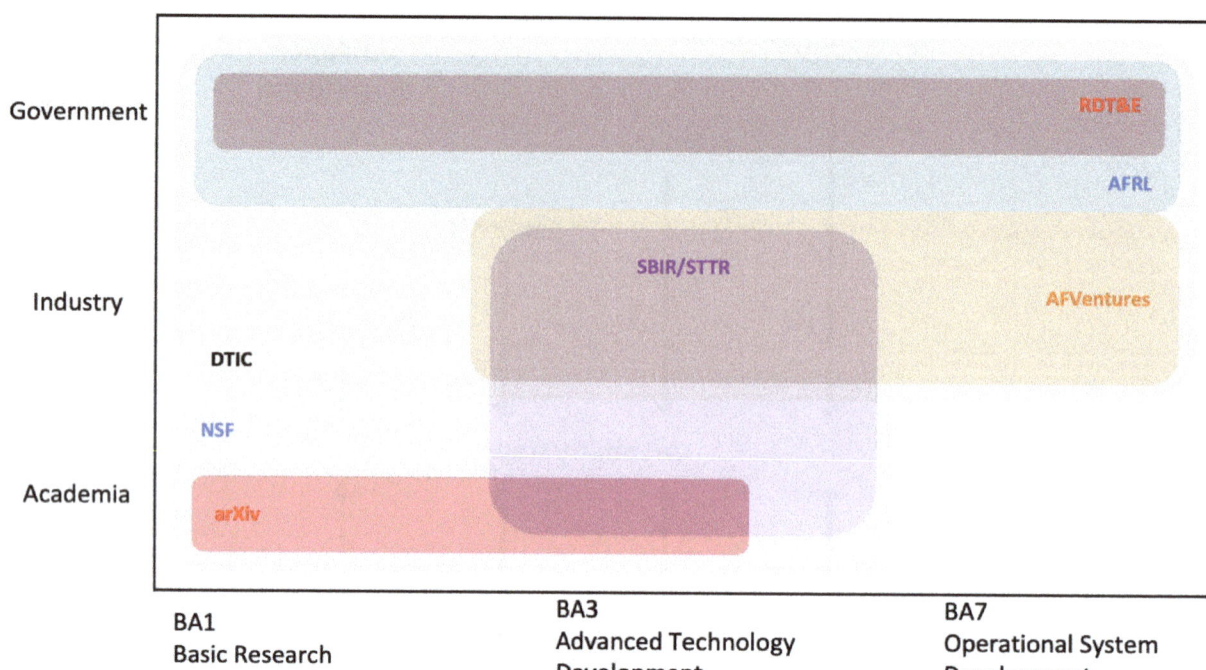

Table 4.1 lists data sources that we compiled. Some sources include operational capability gaps, which describe operational limitations without referring to solutions. Other sources include S&T capability gaps, which describe technical limitations of a system that contribute to an operational capability gap. Still other sources include technical solutions, which describe research and system development activities without referring to an operational or S&T capability gap. These distinctions are not exclusive. Many sources contain a mix of operational capability gaps, S&T capability gaps, and technology solutions. The sources we collected are not exhaustive. However, given their diversity, they allowed us to explore the feasibility of applying a common NLP path to multiple textual data sources.

The data sources have limitations. With respect to capability gaps, CCMD and MAJCOM CGLs and IPLs are authoritative, but they do not follow a uniform template, so the level of detail provided for each entry varies from a single sentence to multiple paragraphs. Additionally, the weight of CCMD needs is implied by the rank ordering of entries within each CGL and IPL. However, the overall significance of each need across all CGLs and IPLs is unknown.[3] Other authoritative sources like the Air Force Strategic Master Plan (AF-SMP) are not routinely updated and may lose currency. Sources like the AF-SMP and the NDS express capability gaps at a level of abstraction that is not directly S&T actionable. Certain JLLIS submissions contain very detailed descriptions of S&T actionable capability gaps, yet the entries are not authoritative. In addition, many JLLIS submissions can be resolved without materiel solutions. Finally, CGL,

---

[3] Needs could be ordered or tiered based on assessments like NDA alignment, risk to mission, or prevalence across commands.

Table 4.1. Data Sources

Source	Description	Capability Gaps	S&T Gaps	Technology Solutions
AFRL FY 2022 POM (unclassified)	Summaries of AFRL research programs	X	X	X
AFVentures portfolio	Company descriptions and key words			X
arXiv abstracts (unclassified)	Abstracts for scholarly scientific publications			X
AF-SMP Capabilities Annex (classified)	List of capability gaps associated with Air Force core functions and grouped into three levels of importance	X	X	
CCMD IPLs (unclassified and classified)	CCMD prioritized capability gaps	X	X	X
DTIC (unclassified)	Repository for research and engineering information			X
Federally funded research and development center reports (unclassified and classified)	Reports that describe a range of military issues related to doctrine, organization, training, materiel, leadership, personnel, facilities, and policy	X	X	X
JLLIS (unclassified and classified)	Repository of warfighter-submitted lessons learned from operations and exercises	X		
MAJCOM capability gaps (unclassified and classified)	MAJCOM prioritized capability gaps	X	X	X
NSF proposals (unclassified)	Abstracts for NSF awards			X
RDT&E defense budget materials (unclassified)	Summaries of DoD-wide RDT&E program elements (Pes) and projects			X
SBIR/STTR awards (unclassified)	Abstracts for SBIR/STTR awards	X		X

IPL, and JLLIS submissions express current capability gaps. If these gaps are not addressed, they will continue to exist in the future. However, these sources do not anticipate new capability gaps that will affect the future force.

With respect to technology solutions, the pedigree of certain data sources is unknown. The arXiv repository contains scholarly articles that have not yet undergone peer review. NSF and SBIR awards contain descriptions of research programs that have been initiated, but the rigor of the awards and the success of the efforts are unknown. Additionally, technology descriptions do not follow a uniform template, so the level of detail can vary greatly. This is particularly challenging in the case of the RDT&E Defense Budget, where a single paragraph may describe a PE or project comprised of ten or more activities valued at hundreds of millions of dollars.[4]

All data sources contain specialized vocabulary that may require subject matter knowledge to interpret. Sources also sometimes use different terms or phrases to refer to the same thing—for

---

[4] In some cases, this problem can be addressed by analyzing the complete text from the submission and not just the summary or abstract.

example, *Joint All-Domain Command and Control* versus *Multi-Domain Command and Control,* or *artificial intelligence* versus *machine intelligence.* Finally, many sources contain repetitive words or phrases that are not relevant to the meaning of the entries themselves. For example, many RDT&E Defense Budget entries contain the phrase "This program element may include necessary civilian pay expenses," and many DTIC entries contain company-specific proprietary notices. These issues can be addressed in the NLP path, whereas limitations related to the completeness, authoritativeness, and pedigree of records remain.

Data Extraction

Data forms varied by source.[5] File types included Microsoft Excel, Word, and PowerPoint; and JSON objects, PDF files, and XML files. All data sources were machine readable, but each required a tailored script to extract the relevant text description and meta tags. For example, Figure 4.4 shows an entry for a PE in the RDT&E Defense Budget. For this data source, the processing script parsed the text in the PDF and extracted the appropriation/budget activity, PE (number/name), cost, and mission description and budget item justification for every unclassified PE and project.

Figure 4.4. Program Element Summary in the Defense Budget

Exhibit R-2, RDT&E Budget Item Justification: PB 2022 Air Force											Date: May 2021	
Appropriation/Budget Activity 3600: Research, Development, Test & Evaluation, Air Force / BA 8: Software and Digital Technology Pilot Programs						R-1 Program Element (Number/Name) PE 0608410F / Air & Space Operations Center (AOC) - Software Pilot Program						
COST ($ in Millions)	Prior Years	FY 2020	FY 2021	FY 2022 Base	FY 2022 OCO	FY 2022 Total	FY 2023	FY 2024	FY 2025	FY 2026	Cost To Complete	Total Cost
Total Program Element	-	0.000	0.000	186.915	0.000	186.915	-	-	-	-	-	-
684596: AOC - Software Pilot Program	-	0.000	0.000	186.915	0.000	186.915	-	-	-	-	-	-
Quantity of RDT&E Articles	-	-	-	-	-	-	-	-	-			

Note
In FY 2022, PE 0608410F, Air & Space Operations Center (AOC), Project 684596, AOC WS Modifications, was transferred from RDT&E, AF, PE 0207410F, Air & Space Operations Center (AOC), Project 674596, AOC WS Modifications, OPAF PE 0207410F, Air & Space Operations Center (AOC), Line Item 834530, AOC, and OMAF Operating Forces, Activity Group, Air Operations, Sub-Activity Group 11C, Combat Enhances Forces, in order to a participate in DoD's Budget Activity 08 (BA08) for Software and Digital Technology Pilot Program.

A. Mission Description and Budget Item Justification
This effort is not a new start. The FY 2018 NDAA Sections 873/874 directed OSD to streamline software development. The Air Operations Center Weapon System (AOC WS) is an OUSD(A&S) pilot initiative in which all lifecycle funding will be tracked under BA08, Software and Digital Technology Pilot Programs. Pilot programs will test the ability to execute modern software development practices encompassing development, procurement, modification and sustainment activities.

The AOC WS, AN/USQ-163 Falconer, the senior element of the Theater Air Control System (TACS), is the weapon system that the Commander, Air Force Forces (COMAFFOR) provides the Combined/Joint Force Air Component Commander (C/JFACC) for planning, executing, and assessing theater-wide air and space operations. The C/JFACC provides air, space and cyber support to the Combined/Joint Forces Commander (C/JFC) by coordinating, deconflicting, and assessing the progress of various weapon systems to advance the C/JFC's campaign. The AOC WS develops operations strategy and planning documents. The weapon system also disseminates tasking orders; executes day-to-day peacetime and combat air, space and cyber operations; and provides rapid reaction to immediate situations by exercising positive control of friendly forces. This program is part of the overarching Kessel Run portfolio.

SOURCE: DoD, 2021, p. 795.

---

[5] CCMD and MAJCOM CGLs and IPLs were available as Excel, PowerPoint, and Word files, and as PDF files. NSF awards were available as XML files, and arXiv abstracts were available as JSON objects. SBIR awards were accessible via a web-based API and were exported as JSON objects. JLLIS records were exported as Microsoft Excel files. Most other data sources (i.e., federally funded research and development center reports, RDT&E Defense Budget Materials, and AF-SMP Annexes) were available as PDF files.

The final database contained a separate table for each source listed in Table 4.1. Each record (e.g., a capability gap from an IPL or an NSF award) appeared as a row in the corresponding table. Records contained free-text data along with meta tags, such as the CCMD or MAJCOM contributing the capability gap or the NSF directorate issuing the award.

Data Preparation and Semantic Analysis

We followed well-established steps for preparing text data for ML, as described in Appendix C. Broadly, these steps involve (1) dividing the raw text from a document into its constituent words and terms; (2) standardizing those words and terms; and (3) computing a measure of the informativeness of each word and term based on its uniqueness to certain documents.

We applied a technique called latent semantic analysis (LSA) to the prepared data (Landauer, Foltz, and Laham, 1998). LSA discovers a set of latent topics present across documents in the data set. For example, a database of capability gaps might contain topics related to command and control (C2) and base defense. Each topic emphasizes certain terms present in the data set. For example, the C2 topic would emphasize terms like *command* and *control*, whereas the base defense topic would emphasize terms like *perimeter* and *missile*. Following LSA, each document is represented as a mixture of topics rather than as a collection of words.

LSA allows the user to do more than just key word searches. By representing documents as mixtures of topics, it becomes possible to compute similarity metrics. This underlies three common uses of LSA:

1. returning documents that are semantically related to a search phrase even if they do not contain the exact terms
2. discovering clusters of semantically related documents
3. returning documents that are semantically related to another document.

We implemented functionality in an application to allow analysts to search and navigate through the databases of capability gaps and technology solutions in these three ways.

Use of SCAnTExT to Enable Natural Language Processing

The NLP path is a foundational element of an HCDE process for deciding which concepts to include in the transformational capability pipeline. Information from the data sources can guide the discovery and selection of capability gaps. For example, the data sources can be used to examine how pervasive a gap is over time and across regions and to characterize the context and conditions that accompany it. In addition, a large cluster of warfighter submissions pertaining to a common operational challenge may signify the need to address that challenge even if it has not yet been identified as a formal gap.

The data sources can also guide the discovery and selection of technology solutions. For example, they can be used to examine how much is being invested in the development of a technology, and in which sectors. In addition, a surge in publications may signify an emerging technology with potential military utility.

Due to limitations in the data sources and the multi-objective nature of forming a research portfolio, the selection of which transformational capabilities to pursue will remain a human-centered process. However, information extracted from data sources using the NLP path can enhance decisions.

The Prototype System

We created the prototype system tool SCAnTExT to allow an analyst to interact with the natural language databases (see Appendix D). Using SCAnTExT, an analyst can perform the three types of semantic operations shown in Figure 4.1: (1) locating records; (2) grouping records; and (3) matching records. To illustrate these semantic operations, we follow the case of an officer in an acquisition career field seeking materiel solutions to improve aeromedical evacuation (AE).

Semantic Operation 1: Locating Records

To demonstrate how SCAnTExT can be used to locate records, we loaded the JLLIS data set. Next we entered a search phrase related to Critical Care Air Transport Teams (CCATTs) and AE. The phrase contained the terms *patient*, *CCATT*, and *AE*. Figure 4.5 shows the table of results returned by SCAnTExT. Records are sorted by semantic similarity to the search phrase. All

Figure 4.5. SCAnTExT Table of Results for JLLIS Search

Title
All
INADEQUATE PREPARATION OF PATIENTS FOR AEROMEDICAL EVACUATION (AE) NORTHCOM
INADEQUATE PREPARATION OF PATIENTS FOR AEROMEDICAL EVACUATION (AE)
Evacuation Patient Visibility
Summary - AE CCATT Liaison located at Lundstuhl Regional Medical Center
Lacking Critical Review of Patient Movement Requests
Aeromedical Evacuation Concept of Operations Not Followed
932 Aeromedical Staging Squadron Scott AFB
Timely notification for CCATT
40 EMDG, 40th Air Expeditionary Wing (AEW), deployed to Diego Garcia
Comments: AFMLO, AFSOC

NOTE: AFB = Air Force Base; AFMLO = Air Force Medical Logistics Office; AFSOC = Air Force Special Operations Command; EMDG = Expeditionary Medical Group.

the top records clearly relate to AE.[6] The user can select and export records from the table. Collectively these records give the acquisition officer examples of AE challenges and operational conditions experienced firsthand by warfighters.

Semantic Operation 2. Group Records

The upper panel of Figure 4.6 shows the visualization generated by SCAnTExT of records in the JLLIS database. Each node in the network diagram is a record. Records that are semantically related have similar colors and are close to one another.[7] Additionally, the pairs of records with highest semantic similarity are linked to one another.[8] The network diagram reveals global structure by clusters of nodes made up of related records. The diagram also reveals local structure by the placement of nodes within each cluster and the links between nodes.

Most search results in Figure 4.5 came from the three clusters highlighted in the lower panel of Figure 4.6. The clusters are related to one another, as evident by their similar colors and the dense interconnections. Yet the records pertain to different facets of AE, so they are placed in different clusters. The user may select an entire cluster from the network diagram and export the records that it contains. This global view gives the acquisition officer a sense of the scope and magnitude of AE-related challenges.

Semantic Operation 3. Link Records

The upper panel of Figure 4.7 shows a visualization generated by SCAnTExT of records in the unclassified CCMD IPL database. Once again, the placement of nodes reveals global structure (i.e., clusters of semantically related nodes) and local structure (i.e., the placement of nodes within clusters and the links between nodes). The figure contains distinguishing words from three clusters to illustrate some of the themes present in the data: *interoperable communications*, *defeat of unmanned aerial systems* (UAS), and *directed-energy/lasers*.

SCAnTExT can also link records between data sets. The node placed above the network diagram in Figure 4.7 (bottom panel) contains text from the 96 records in the JLLIS CCATT cluster. This search node is connected to the two IPL records that are most semantically related to it: (1) *extend the "golden hour,"*[9] and (2) *provide viable solutions for multimodal patient movement.* As this example illustrates, SCAnTExT can be used to automatically align informal warfighter submissions contained in one database to formal capability gaps contained in another. In this way, the acquisition officer can build a more comprehensive case—traceable to warfighter experiences and DAF guidance—for new AE capabilities.

---

[6] Only the top ten results are shown. Many more results are related to AE and can be accessed using the tool.

[7] We performed a hierarchical cluster analysis to define groups of related records. For each data set, we experimented with different numbers of clusters to determine settings that returned relatively homogeneous groups of records.

[8] Specifically, we linked each record with its single nearest neighbor.

[9] The "golden hour" refers to the period of time following a traumatic injury during which there is the highest likelihood that medical care will prevent death.

Figure 4.6. Network Visualization of JLLIS Data

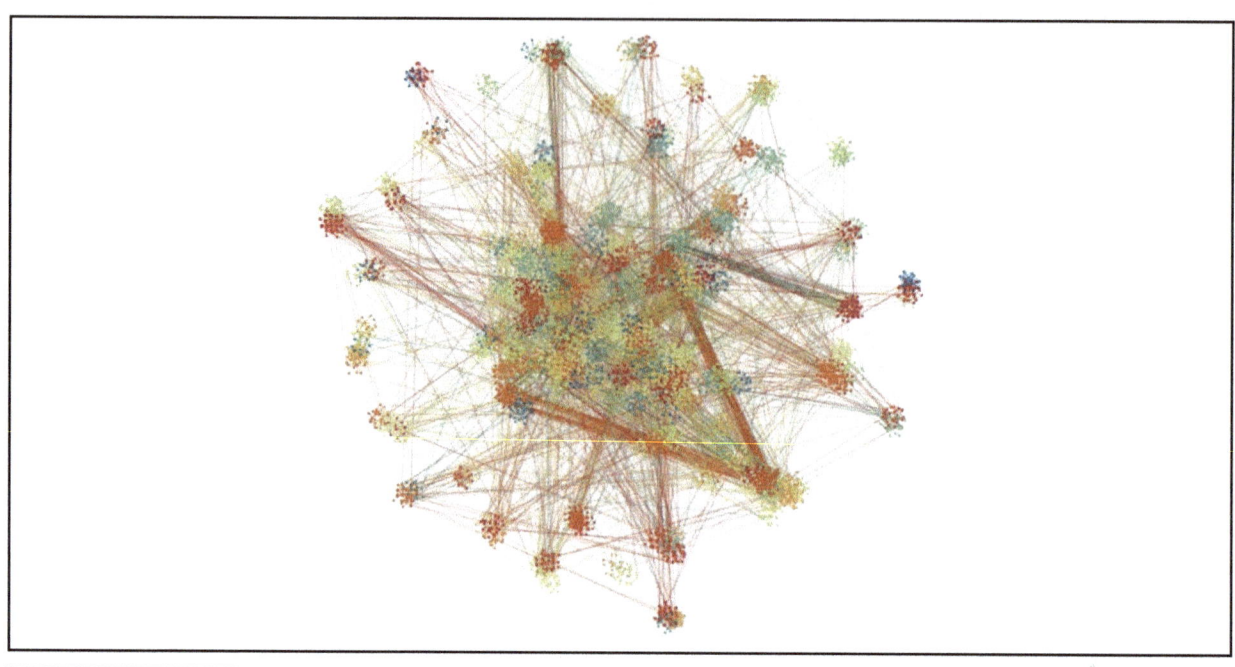

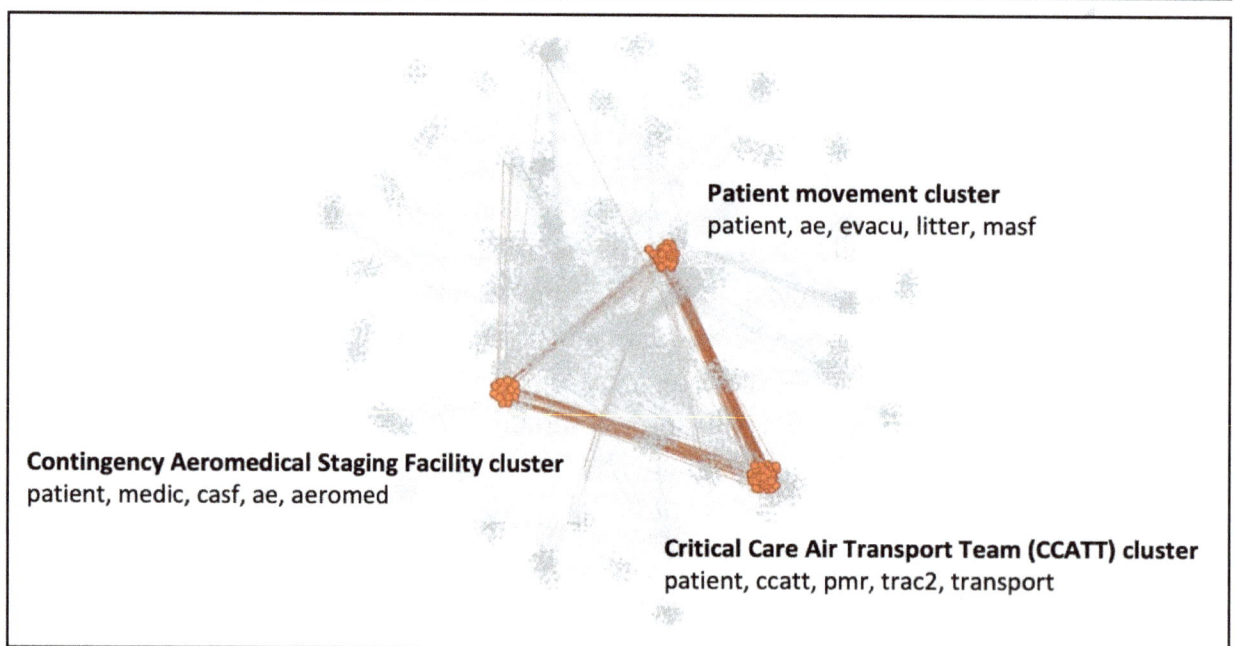

Patient movement cluster
patient, ae, evacu, litter, masf

Contingency Aeromedical Staging Facility cluster
patient, medic, casf, ae, aeromed

Critical Care Air Transport Team (CCATT) cluster
patient, ccatt, pmr, trac2, transport

NOTE: The lists below cluster titles show the highest frequency tokens within each cluster. Some, like casf are standard abbreviations (e.g., contingency aeromedical staging facility). Others are words (e.g., patient) or word stems (e.g., evacu).

Figure 4.7. Network Visualization of Unclassified Combatant Command Integrated Priority Lists

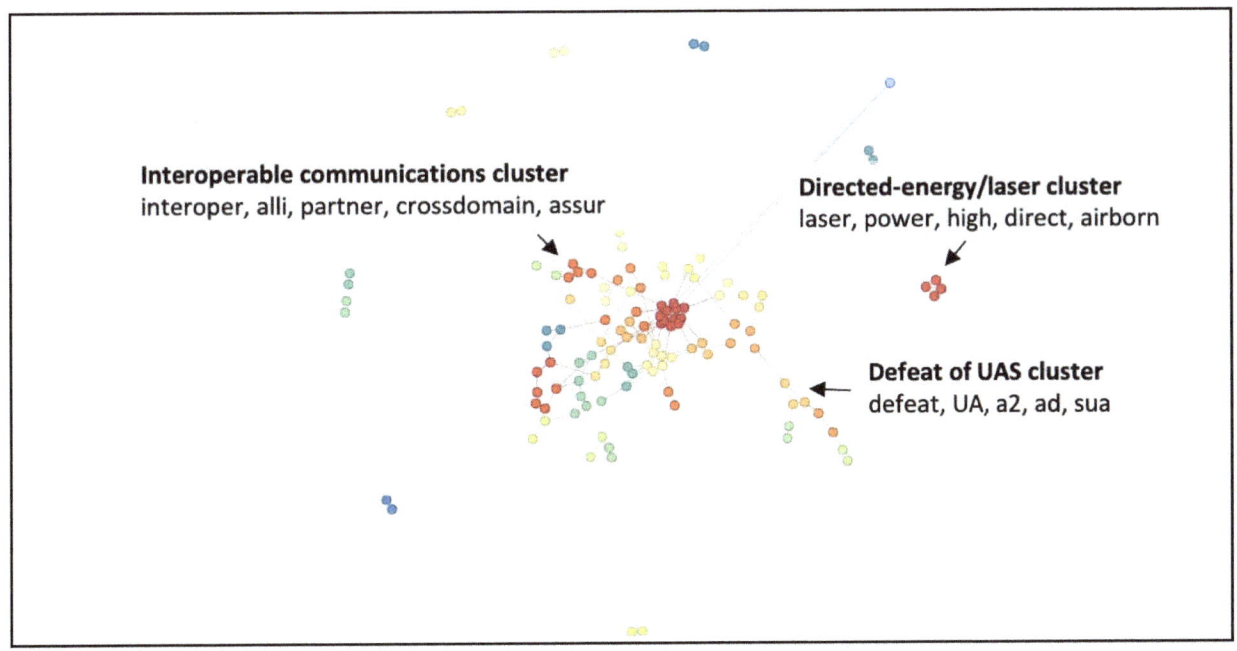

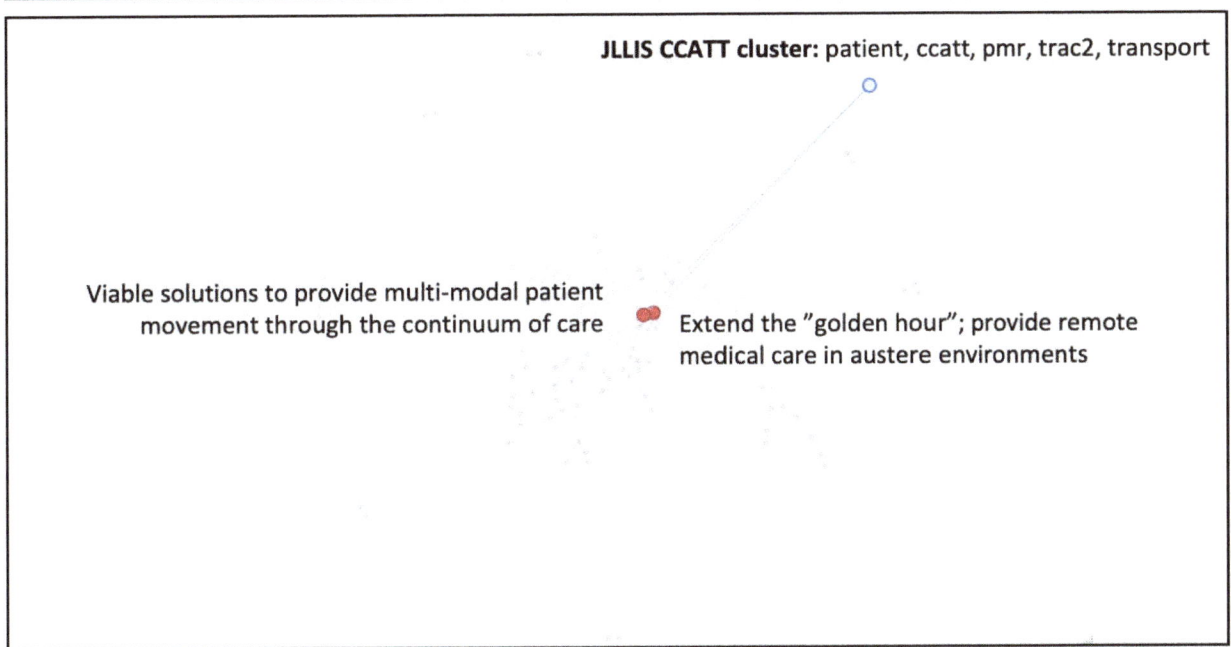

NOTE: The lists below cluster titles show the highest frequency tokens within each cluster. Some, like ccatt, are standard abbreviations (e.g., critical care air transport team). Others are words (e.g., patient) or word stems (e.g., evacu).

The upper panel of Figure 4.8 shows a visualization generated by SCAnTExT of records contained in the AFVentures database. The figure contains distinguishing terms from three clusters to illustrate some of themes present in the data: *quantum computing, circuits and semiconductors*, and *medical.*

Figure 4.8. Network Visualization of AFVentures Company Profiles

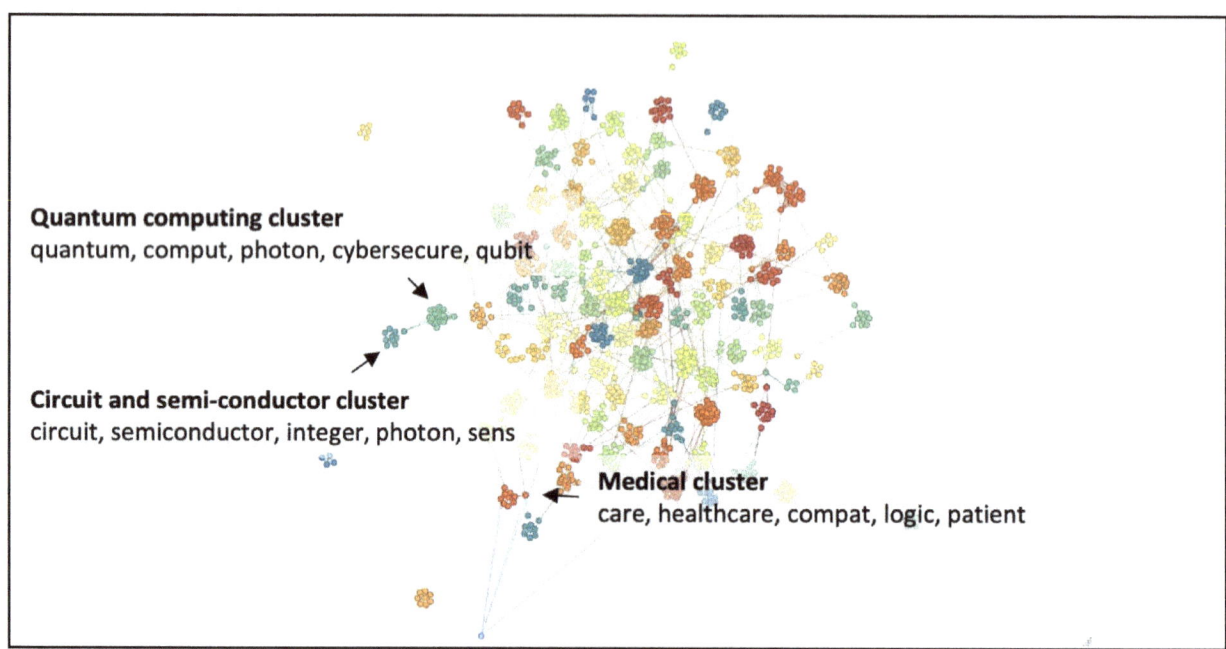

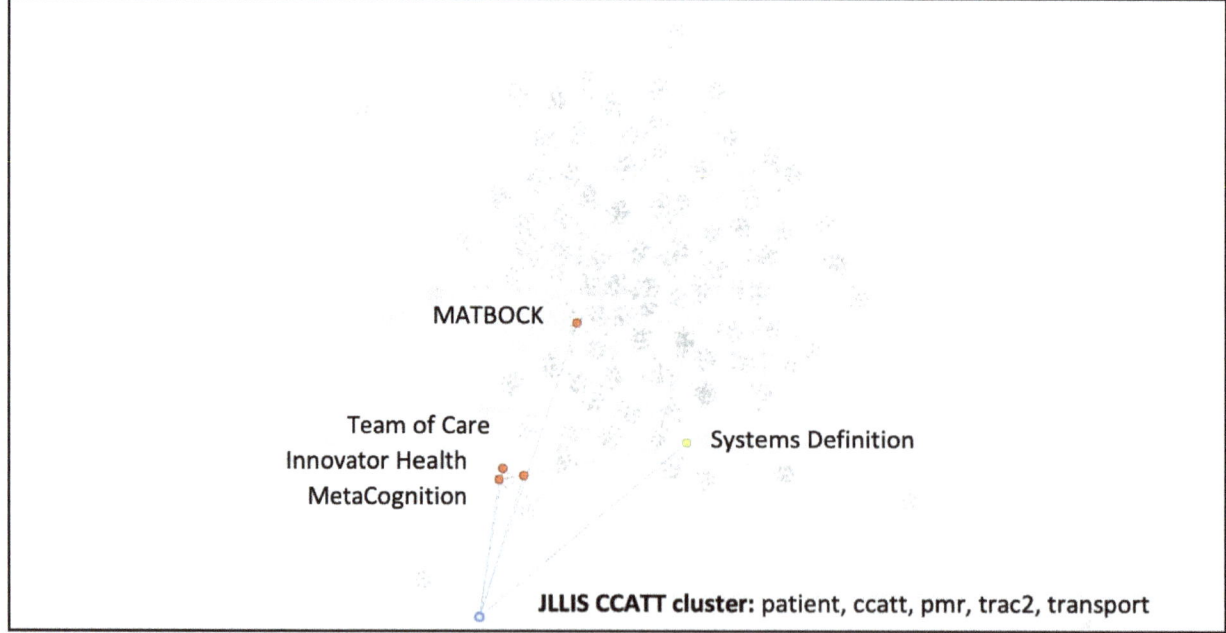

NOTE: The lists below cluster titles show the highest frequency tokens within each cluster. Some, like ccatt are standard abbreviations (e.g., critical care air transport team). Others are words (e.g., patient) or word stems (e.g., evacu).

The node placed below the network diagram in the lower panel of Figure 4.8 contains text from the 96 records in the JLLIS CCATT cluster. The search node is connected to five AFVentures records that are most semantically related. The names of the five companies are Innovator Health, MATBOCK, MetaCognition, Systems Definition, and Team of Care. The profiles for the companies describe business offerings in emergency medicine, AE, and transport. As this example illustrates, SCAnTExT can be used to automatically link warfighter capability gaps to technology solutions and companies that offer them. By using this functionality, the acquisition officer can identify businesses that may be able to address AE capability needs.

Together these examples show how an acquisition officer, or anyone working in a related area, can use SCAnTExT to identify clusters of warfighter needs, collate them with formal capability gaps, and use them to search for technology solutions. Although the search functionality is basic, few of the data sources contained in Table 4.1 come with an interface that allows a user to perform a search. All-purpose tools like Microsoft Excel exist, but they rely on exact matches rather than semantic similarity to search terms and phrases, and so they are more brittle. Additionally, none of the data sources contains an interface that allows the user to visually scan the complete database to identify clusters of related records. Finally, none of the data sources allows users to simultaneously search among different databases.

Chapter Summary

In this chapter we have described an NLP path to prepare text data for NLP and apply semantic operations to extract information from text data. The data sources and the NLP methods have limitations. Yet it was possible to locate records, group records, and match records based on semantic similarity. The prototype system for applying the NLP methods to the various data sets constitutes a critical element of an HCDE process for deciding which concepts to feed into the transformational capability pipeline. The prototype is complementary to other systems designed to track technology developments in DoD, but it goes beyond those systems by also capturing technology developments in industry and academia, along with operational capability gaps.

One of the issues raised during interviews was that capability gap and technology scans can be incomplete and potentially biased. By combining records from multiple sources, SCAnTExT lessens, but does not formally eliminate, these concerns. The data sources only cover part of the operational and technology ecosystem, and the ultimate results depend on which outputs the user accepts from SCAnTExT. User testing is needed to mature the proof of concept and to operationalize its use. In addition, to evaluate the prototype, we considered the number and diversity of relevant records retrieved. Further testing is needed to demonstrate the effectiveness of the tool in enhancing development planning.

Chapter 5. Foresight Methods to Support an HCDE Decision Process

A key element of an HCDE process is to use human intelligence to synthesize information from data sources into comprehensive descriptions of capability gaps and corresponding technology solutions (i.e., a human-centered decision process). In this chapter we consider the suitability of a class of foresight methods to support the human dimension of HCDE decision processes.

Future-oriented technology analysis is challenging for two reasons: uncertainty about the pace and nature of technology developments and uncertainty about the future global environment and the resulting operational needs (Leftwich et al., 2019). These challenges are not unique to the TCO. Other communities in academia, industry, and government must also perform future-oriented analyses (Keenan, Barré, and Cagnin, 2008, p. 170). These communities use a set of techniques—collectively called foresight methods—to perform future-oriented analyses. Foresight methods use critical thinking, planning, and management competencies to account for long-term uncertainties when making short-term decisions (Greenblott et al., 2019). Foresight methods are intended to increase understanding of the complex dynamic systems that shape future technology opportunities and needs, and they engage multiple stakeholders from different backgrounds (Ye and Feng, 2013). Given the TCO's mission to prioritize, demonstrate, and transition capabilities for future air and space forces, the TCO may also benefit from employing foresight methods in its future-oriented analyses.

Foresight brings direct value to organizations by allowing them to expand into new markets, discover innovative concepts, and increase the quality of the outputs of innovative applications (Rohrbeck and Gemünden, 2010). For businesses this translates to increased revenue or, in the case of deferred or canceled initiatives, increased savings. For the TCO this translates to increased warfighter effectiveness through the successful maturation of concepts that deliver critical capabilities.

Foresight brings indirect value to organizations, as well (Boe-Lillegraven and Monterde, 2015), including the following:

- Individual cognitive benefits: Participation in foresight activities improves an individual's mental model of a particular topic and encourages critical and creative thinking. This would benefit the TCO by enhancing people's understanding of complex problems and by exposing them to new decisionmaking frameworks.[1]

---

[1] These benefits would extend to participants from S&T, operational, and acquisition communities, along with offices that play an integrating role, such as the AFRL Plans and Programs Directorate and the TCO.

- Collective cognitive benefits: Use of foresight activities by an organization increases shared understanding of a particular topic and encourages divergent thinking among individuals. This would benefit the TCO by increasing understanding across the large set of DAF stakeholders and evoking different perspectives.
- Collective consensus: By tapping into a broad set of values that reflect many individual interests, foresight activities increase consensus and adoption (Schatzmann, Schäefer, and Eichelbaum, 2013). This would benefit the TCO by pointing S&T, operational, and acquisition communities toward a common objective.

Next we review foresight methods and conditions for their use. A key question in this regard is which methods to use for a given problem. To demonstrate how to select which foresight methods to include in an HCDE process, we evaluate three TCO initiatives—Explore, WARTECH, and Vanguard—and we recommend forecasting packages, or "playbooks," for each.

Foresight Methods

Foresight methods involve sharing a vision and/or a set of objectives among key stakeholders, including SMEs, operators, and decisionmakers. Foresight methods typically draw from interdisciplinary research fields, and they encompass combinations of quantitative and qualitative research methodologies.

Attributes of Foresight Methods

The **nature** of foresight methods can be categorized as *qualitative*, *quantitative*, or *semiquantitative* (Popper, 2008). Qualitative methods rely on subjective but informed interpretations and perceptions of events. Quantitative methods rely on repeatable, objective measurement and analysis. Semiquantitative methods rely on mathematical principles to quantify subjectivity such as weighting of opinions (Popper, 2008). The **capabilities** of foresight methods can be characterized as a blend of methods that are *evidence based*, *expertise based*, *interactive*, and *creative*. Evidence-based methods explain and/or forecast a particular phenomenon with the support of reliable documentation and means of analysis; expertise is associated with the skills and knowledge of individuals in a particular area or subject; interactive refers to gathering informed individuals to articulate their viewpoints; and creativity is the mixture of original and imaginative thinking. These attributes are the building blocks of the "foresight diamond" (Figure 5.1), which here shows the 33 foresight methods according to their attributes.[2] In our

---

[2] Table B.1 describes the 33 foresight methods shown in the diamond.

Figure 5.1. The Foresight Diamond

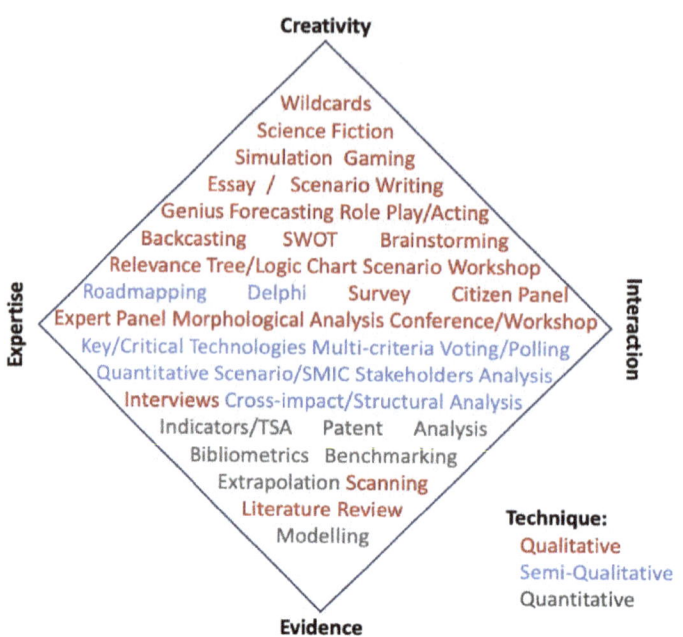

SOURCE: Popper, 2008.
NOTE: SMIC = Systèmes et Matrices d'Impacts Croisés; SWOT = strengths, weaknesses, opportunities, and threats; TSA = time series analysis.

interviews we heard about limited use of some of these methods: brainstorming, scenarios, conferences/workshops, scenarios, horizon scanning, literature review, and modeling.

Conditions for Foresight Methods

A key question for an organization is what foresight method or mixture of methods is most suitable for its needs. Methods that are not suitable may lead to shallow insights or may not be executable given the organization's constraints.

Porter (2010) established nine dimensions to differentiate between different types of foresight problems (see Table 5.1). Each dimension has multiple state values. The state values along the nine dimensions can be used to determine the suitability of different foresight methods for a given problem or program.

Table 5.1. Dimensions and State Values of Foresight Problems

Dimensions	State Values			
Motivation	Open-ended	Directed		
Drivers	Science	Technology	Innovation	Context
Scope	Single technology	Multiple technologies	Wide-ranging planning	
Locus	Institution	Sector	National	Global
Time horizon	Short (1–2 years)	Mid-range (3–10 years)	Long (10+ years)	
Purpose	Informational	Action oriented		
Target users	Few	Diverse		
Participation	Narrow	Intermediate	Diverse	
Study duration	Day(s)	Month(s)	Year(s)	

SOURCE: Adapted from Porter, 2010; minor changes to state-value nomenclature and timeframes made by the authors of this report, but Porter's intent is maintained.

Determining Conditions for Transformational Capabilities Office Initiatives

To determine the suitability of different foresight methods, we applied Porter's framework to the processes underlying three TCO initiatives: Explore, WARTECH, and Vanguard (see Figure 5.2).[3] The *motivation* for a problem may be open-ended, meaning the intended future state is not yet defined, or it may be directed, meaning that the organization is trying to achieve an agreed-upon future objective. All TCO initiatives are directed, but some Explore challenges may also be considered open-ended.

The *drivers* for a problem may be consideration of changes in any of the following: science, or basic research; technology, or applied research and engineering; innovation, or novel uses of technologies; and context, or a change in the environment. All TCO initiatives consider technology and innovation drivers. Some Explore challenges have a longer time horizon and so may consider science drivers. WARTECH also considers changes to context as a driver.

The *scope* of a problem may be limited to a single technology, it may encompass multiple technologies, or it may entail wide-ranging planning (to include multiple technologies). Each Explore challenge tends to emphasize one technology, whereas Vanguards incorporate multiple technologies (i.e., systems of systems), and WARTECH involves wide-ranging planning.

The *locus* of a problem may be an institution such as the Air Force, a sector such as the DoD and the defense industry, a nation, or multiple nations. The locus for all TCO initiatives includes

---

[3] To determine state values along the nine dimensions for Explore, WARTECH, and Vanguard, we reviewed source material about each initiative, and we conducted discussions with individuals involved in each initiative. The characteristics of a process may differ from the characteristics of its outcome. For example, each Explore challenge is targeted, yet because the challenges are scientific in nature, they may have broadly applicable outcomes. When evaluating the nine dimensions, we considered the processes, rather than the potential outcomes, of the different initiatives.

Figure 5.2. State Values for Transformational Capabilities Office Initiatives

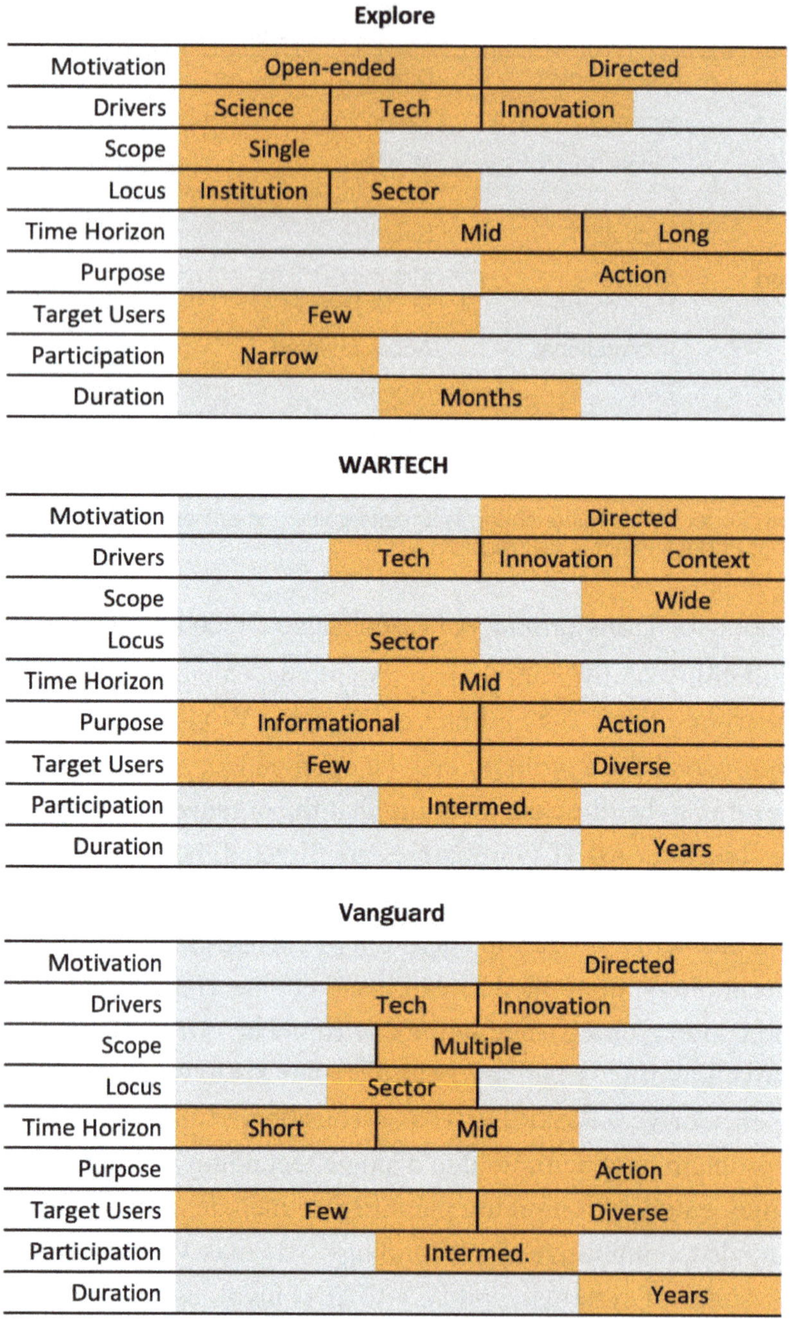

Explore

Motivation	Open-ended		Directed	
Drivers	Science	Tech	Innovation	
Scope	Single			
Locus	Institution	Sector		
Time Horizon			Mid	Long
Purpose			Action	
Target Users	Few			
Participation	Narrow			
Duration		Months		

WARTECH

Motivation			Directed	
Drivers		Tech	Innovation	Context
Scope				Wide
Locus		Sector		
Time Horizon			Mid	
Purpose	Informational		Action	
Target Users	Few		Diverse	
Participation			Intermed.	
Duration				Years

Vanguard

Motivation			Directed	
Drivers		Tech	Innovation	
Scope			Multiple	
Locus		Sector		
Time Horizon	Short	Mid		
Purpose			Action	
Target Users	Few		Diverse	
Participation			Intermed.	
Duration				Years

the DoD and the defense industry, a sector. The locus for some Explore challenges more narrowly encompasses the Air Force, an institution.

The *time horizon* of a problem may be short (one to two years), medium (three to ten years) or long (ten or more years). Explore challenges tend to have a medium or long horizon, WARTECH has a medium horizon, and Vanguard has a short horizon.

The *purpose* of a foresight exercise may be either to generate information, such as understanding the relationships between variables in a dynamic system, and their effects on future states; or to drive action, such as allocating funding and performing RDT&E activities. The purpose of all TCO initiatives is to drive action; however, an additional aim of WARTECH is to generate information.

There are few *target users* for a future technology if the technology is intended for one functional or operational community; there are diverse target users if the technology is intended for multiple communities. Explore challenges have a relatively narrow scope and so involve few different target users, whereas WARTECH and Vanguard concern crosscutting capabilities intended for diverse target users.

Participation in foresight may be narrow (i.e., individuals from the same background), intermediate (i.e., individuals from different backgrounds), or diverse (i.e., individuals from all backgrounds representative of target users). Participation in Explore tends to be narrow relative to participation in WARTECH and Vanguard.

Finally, the *duration* of foresight activities may span days, months, or years. A typical Explore challenge lasts from several months to a year, whereas WARTECH and Vanguard may extend multiple years.

The selection of foresight methods depends on other factors like funding and classification level. Factors such as these may limit a study's scope, participation, and duration, which has implications for the suitability of different methods.

Determining Conditions for Foresight Methods

After applying Porter's framework to TCO initiatives, we then applied it to the 33 foresight methods listed in the foresight diamond (Figure 5.1) and defined in Appendix B. We used criteria described by Porter to determine which state values suited each foresight method. To illustrate, we describe three representative methods: (1) science fictioning incorporates imaginative, plausible content, and is based in scientific facts, theories, and principles; (2) modeling involves building a mathematical or computational representation of a process to simulate its outcomes; and (3) backcasting starts from a desired future and works backward to identify efforts, policies, and programs to navigate toward that goal from the current state.

Figure 5.3 shows the three methods with their respective state values.[4] Most methods were applicable given many, but not all, state values. For example, science fiction can be used to explore open-ended futures (*motivation*), it can be applied to long timescales (*time horizon*), and it can generate innovative uses of S&T (*drivers*). Modeling can be used to explore the utility of a single or limited set of solution options (*scope*) enabled by advances in science or technology (*drivers*) and applicable across short and medium timescales (*time horizon*). Finally, backcasting can be used to work backward from a directed end state (*motivation*), it can incorporate all manner of *drivers*, and it can account for wide-range planning that goes beyond just technology solutions (*scope*).

---

[4] Table B.2 shows complete mapping of conditions to foresight methods.

Figure 5.3. State Values for Select Foresight Methods

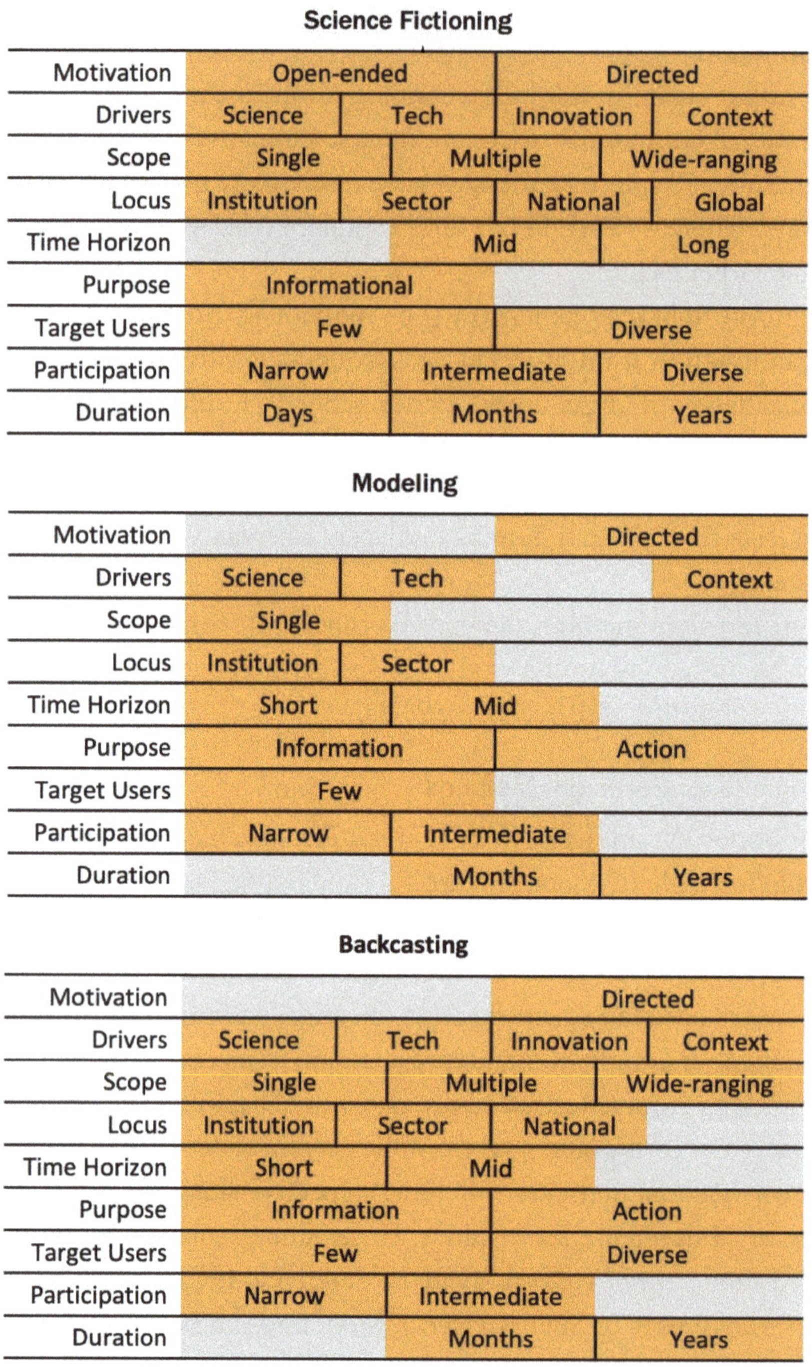

Science Fictioning

Motivation	Open-ended		Directed	
Drivers	Science	Tech	Innovation	Context
Scope	Single	Multiple		Wide-ranging
Locus	Institution	Sector	National	Global
Time Horizon			Mid	Long
Purpose	Informational			
Target Users	Few		Diverse	
Participation	Narrow	Intermediate		Diverse
Duration	Days	Months		Years

Modeling

Motivation			Directed	
Drivers	Science	Tech		Context
Scope	Single			
Locus	Institution	Sector		
Time Horizon	Short	Mid		
Purpose	Information		Action	
Target Users	Few			
Participation	Narrow	Intermediate		
Duration		Months	Years	

Backcasting

Dimension				
Motivation			Directed	
Drivers	Science	Tech	Innovation	Context
Scope	Single	Multiple		Wide-ranging
Locus	Institution	Sector	National	
Time Horizon	Short	Mid		
Purpose	Information		Action	
Target Users	Few		Diverse	
Participation	Narrow	Intermediate		
Duration		Months	Years	

Determining the Suitability of Foresight Methods for Transformational Capabilities Office Initiatives Based on the Conditions

We determined which foresight methods best aligned with each of the TCO initiatives based on the nine dimensions. For example, the *purpose* of science fiction is to generate information, so it is partially aligned with WARTECH for that dimension, but it is not aligned with Explore or Vanguard. For all combinations of TCO initiative and foresight method, we computed the percentage of dimensions where the state value given for the TCO initiative was included in the set of state values that the foresight method was most suitable for.[5] A value of 100 percent indicates that the foresight method is suitable for the TCO initiative for all nine dimensions, and a value of 0 percent indicates that it is not suitable for any dimension. To return to the example from above: the percentage overlap between Explore and science fiction is 88 percent because the two are aligned for eight dimensions and they are not aligned for one dimension (i.e., *purpose*); the percentage of overlap between WARTECH and science fiction is 93 percent because the two are aligned for eight dimensions and they are partially aligned for one dimension (i.e., *purpose*); and the percentage of overlap between Vanguard and science fiction is 83 percent because they are aligned for six dimensions, they are partially aligned for one dimension (i.e., *time horizon*), and they are not aligned for one dimension (i.e., *purpose*).

Playbook of Foresight Methods for the Transformational Capabilities Office

Table 5.2 shows the ten foresight methods that are most suitable for each of the three TCO initiatives. Some methods are suitable for all initiatives (i.e., brainstorming and science fiction), some are suitable for two initiatives (e.g., backcasting), and some are suitable for one initiative (e.g., scanning). Overall, seven of the ten methods overlap between WARTECH and Vanguard, whereas fewer methods overlap between Explore and WARTECH or Vanguard. This is consistent with the fact that WARTECH generates concepts that may become Vanguards, and in this way, Vanguard can be seen as an extension of WARTECH.

Table 5.2. Applicability of Foresight Methods to Transformational Capabilities Office Initiatives

Method	Explore	WARTECH	Vanguard
Brainstorming	X	X	X
Science Fiction	X	X	X
Backcasting		X	X
Multicriteria analysis		X	X
SWOT		X	X
Structural analysis	X		
Relevance trees		X	X
Quantitative scenarios	X		X
Morphological analysis			X
Scenarios		X	X
Modeling	X		

---

[5] If a TCO initiative had two or more values for a state dimension (e.g., open-ended *and* directed motivation for Explore), we computed the percentage of state values for that dimension that were included in the set for a given foresight method.

Method	Explore	WARTECH	Vanguard
Scanning	X		
Conferences		X	X
Surveys		X	
Roadmapping	X		
Delphi method		X	
Literature review	X		
Benchmarking	X		
Wild cards	X		

Figure 5.4 overlays the "playbooks" of methods for each TCO initiative on the foresight diamond. Two methods were suitable for all TCO initiatives: Brainstorming and science fiction. Brainstorming (i.e., using group working sessions to generate new ideas around a specific area of interest) is so flexible that it is suitable for nearly all foresight endeavors, while science fiction addresses commonalities across these three TCO initiatives.

Figure 5.4. Foresight Packages for Transformational Capabilities Office Initiatives

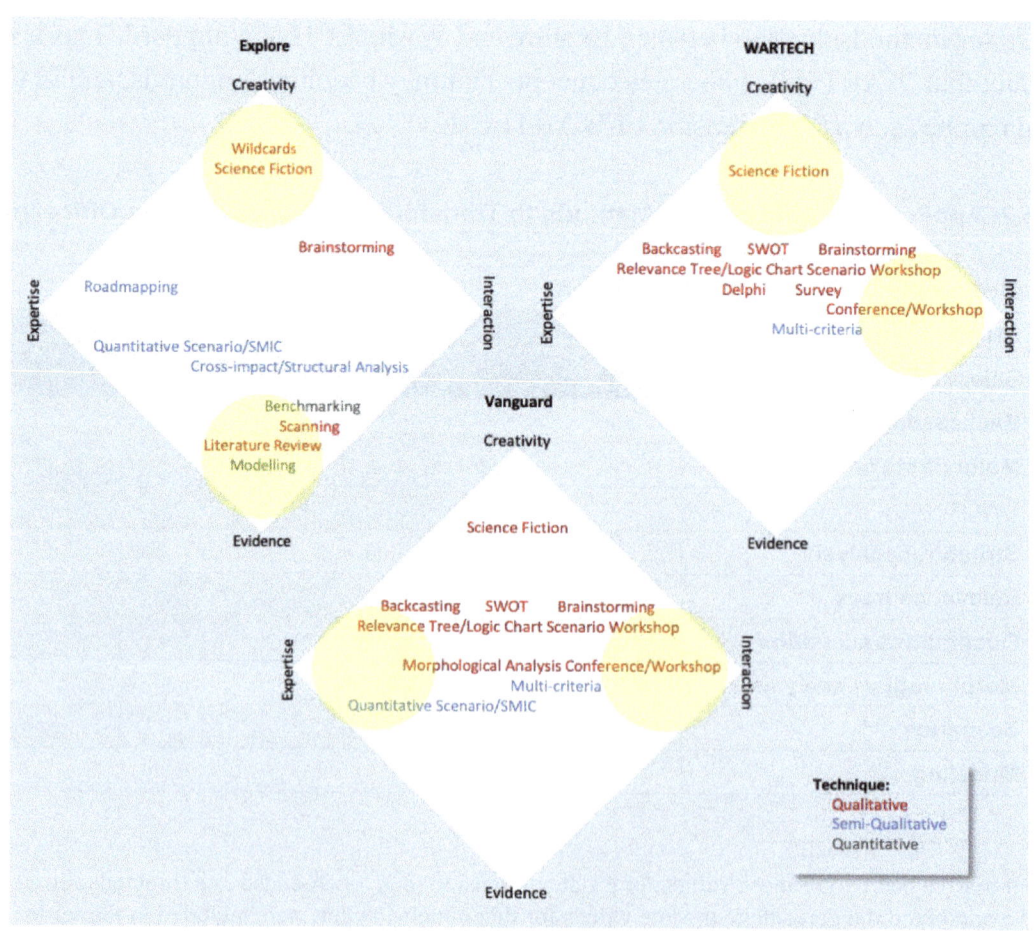

The remaining methods were particularly beneficial for some, but not all, TCO initiatives. The following differences are traceable to the dimensions and state values for each initiative:

- Explore: Creative methods are favored because of the longer horizon under consideration and the open-ended nature of some challenge problems. Evidence-driven methods are also favored because of the narrower scope of challenge problems.
- WARTECH: Creative methods are favored because of the diverse set of drivers and wide scope. Interaction-based methods are also favored because of the need to capture the different values held by diverse target users and to build consensus.
- Vanguard: Interaction-based methods are favored because of the need to account for the different values held by diverse target users and to build consensus. Expertise-driven methods are also favored because of the limited range of technology solutions under consideration.

The combination of foresight methods—or the "methods mix"—is beneficial for drawing out insights that would otherwise be missed using a single method or class of methods (Popper, 2008). For this reason it is important to use methods from different quadrants of the foresight diamond. For example, Explore could begin by *scanning* to identify potential high-leverage technologies that are just beginning to emerge, followed by *modeling* to demonstrate operational utility of those technologies in different CONOPs, followed by *wild cards* to identify events or developments that would dramatically alter the need for, or feasibility of, a particular technology. WARTECH could begin by *brainstorming* to generate ideas for new operational concepts, followed by *conferences/workshops* to refine and socialize concepts with diverse stakeholders, followed by *multicriteria analysis* to determine the extent to which technology enabled CONOPs address myriad operational needs. Finally, Vanguard could use *multicriteria analysis* to identify key system performance attributes, followed by *quantitative scenarios* to demonstrate and experiment with the system, followed by *backcasting* to identify the complete set of doctrine, organization, training, materiel, leadership, personnel, facilities, and policy steps needed to arrive at the intended future state.

The methods shown in Figure 5.4 are not meant to be exhaustive. For example, if a Vanguard project is decomposed into separate elements, modeling may become highly applicable. The methods shown in the figure serve as a starting point and illustrate options that are typically overlooked (e.g., science fiction).

Integration with SCAnTExT

A data science tool like the one described in Chapter 4 can be used to implement certain quantitative, evidence-based foresight methods—for example, literature review and scanning. The tool can also enable the implementation of methods from other quadrants of the foresight diamond; for example, by identifying academic/industry leaders to include on an expert panel, by bounding operational conditions to include in scenario gaming, or to analyze stakeholders' (i.e., MAJCOM's or functional communities') needs.

Chapter Summary

Future-oriented technology analysis presents unique and complex challenges. However, the selection of foresight methods across TCO initiatives, as demonstrated in this chapter, offers a way forward for the Air Force. Our findings underscore both the similarities and differences across the Explore, WARTECH, and Vanguard initiatives. While the three initiatives share certain state values—notably, motivation, target users, drivers, and time horizon—they also have clear differences that suggest divergent foresight methods. By considering alternative, diverse foresight methods, one discards the notion that foresight is a "singular activity with a 'one size fits all' methodology" (Porter, 2010, p. 43). The vast number of foresight methods can offer new, multidimensional observations beyond traditional approaches.

Chapter 6. Case Studies

To illustrate how the data science tool (SCAnTExT) described in Chapter 4 and the foresight methods described in Chapter 5 may be used together to help identify capability gaps and technology solutions, we conducted three case studies. Each case study follows the capability development decisionmaking process shown in Figure 3.1.

The process has four major steps: (1) Use SCAnTExT to extract an initial set of capability gaps from both formal guidance and documents reflecting operational experience; (2) use foresight methods to iterate this selection of gaps to produce a final set, with more detail and more complete descriptions, drawing on SCAnTExT as needed; (3) use SCAnTExT again to extract an initial set of potential technology solutions from various S&T sources, depending on the need; and (4) reiterate with (different) foresight methods to determine the final technology solution set to be proposed.

Note that each case study in this chapter addresses only a portion of this larger process; for obvious reasons, we do not attempt to work through the entire capability development decisionmaking process here.

Case Study 1. High-Speed Vertical Takeoff and Landing

The Air Force and U.S. Special Operations Command require an HSVTOL aircraft with increased range, survivability, and payload to replace the CV-22 Osprey. This requirement is driven by myriad factors, including the need to perform infiltration, exfiltration, personnel recovery (PR), and AE operations in regions that span great distances (e.g., U.S. Indo-Pacific Command); the need to increase aircraft survivability in highly contested environments; and the need to operate in austere conditions and/or places without runways. In recognition of the importance of HSVTOL, AFWERX recently designated an HSVTOL Challenge in conjunction with TCO, launching an HSVTOL Explore 2.0 topic.[1]

In this case study, we began by using SCAnTExT to trace the need for HSVTOL to formal capability gaps and informal warfighter submissions about operational limitations. We then applied a foresight method, brainstorming, to a pair of CONOPs created by AFSOC to identify primary and secondary HSVTOL technical capability needs. Finally, we used SCAnTExT to identify relevant technology developments across military and commercial sectors.

---

[1] AFWERX, HSVTOL Concept Challenge, webpage, undated.

Step 1. Extract an Initial Capability Gap Set

The AFWERX HSVTOL Challenge presents four mission profiles based on infiltration/exfiltration of Special Operations Forces, tactical mobility (i.e., runway-independent tactical airlift), PR, and AE. We searched the formal capability gaps documents and the informal warfighter submissions based on the four CONOPs. Table 6.1 shows alignment between the CONOPs and needs expressed in the data sources.[2] AE appears most consistently, but this is only because the data sources included two Capabilities Based Assessments (CBAs) by medical organizations. Tactical mobility and PR also appeared in most sources.

Table 6.1. Alignment Between High-Speed Vertical Takeoff and Landing Mission Profiles and Data Sources

Source	Infiltration/Exfiltration	Tactical Mobility	PR	AE
Air Force Universal Task List (AFUTL)	X	X	X	X
Command IPLs		X	X	X
CBA for health services support to agile combat employment CONOPS in PACAF				X
JLLIS		X	X	X

The data sources do not directly call for HSVTOL, which is a technical capability. Rather, they identify operational needs that HSVTOL could conceivably address. For example, one unclassified CCMD capability need calls for "automated loading/offloading systems; rapid distribution technologies; innovative delivery technologies; rapidly establish ports of debarkation in austere/anti-access environments." HSVTOL could address this need through its combination of speed, capacity, and runway independence. Surprisingly, our search for tactical mobility also led to records on runway repair.[3] The operational challenge in these cases—runway damage—could be addressed by repair technologies as identified in those records. Alternatively, the operational challenge could be addressed by HSVTOL.

Step 2. Iterate the Selection of Gaps

AFSOC created two HSVTOL CONOPs (AFSOC, September 15, 2020). The first involves performing influence operations by delivering supplies to partner nationalists in a contested territory (i.e., tactical mobility), and the second involves performing PR in a contested open

---

[2] We used the descriptions of the CONOPs as inputs to search the capability gaps.

[3] Although the search phrase for tactical mobility did not contain words like *runway*, *damage*, or *repair*, records on runway damage and repair were nonetheless semantically related to the search phrase, and so appeared in the search.

ocean environment.[4] In both CONOPs, vertical takeoff and landing is needed due to the lack of runways, and high speed is needed to allow the aircraft to link with fighter escorts to increase survivability in a contested environment. A group of SMEs reviewed the scenarios and identified a partial set of key technologies for each (see Table 6.2). Several of these technologies are also directly identified in the AFWERX HSVTOL Challenge and the AFSOC CONOPs. These encompass primary technical capabilities related to structure, aerodynamics, and propulsion, along with secondary technical capabilities. The relevance of the capabilities to the two AFSOC CONOPs is denoted in the table.

Table 6.2. High-Speed Vertical Takeoff and Landing Technologies

Technology	CONOPs		Service RDT&E			
	Mobility	PR	U.S. Army	U.S. Navy	Joint	DTIC
Horizontal high-speed flight	X	X	X			X
Hovering and vertical takeoff and landing	X	X	X			X
Airframe/chassis designs and structures	X	X	X			X
Benign downwash		X				X
Articulating landing gear for all types of terrain						X
Threat detection and self-protection		X	X	X		X
Signature management			X			X
Optional manning			X			X
Ship-based takeoff and landing	X					X
Night vision compatible	X	X	X			X
Terrain following	X		X			X
Joint interoperable communications	X	X	X		X	X
Fourth- and fifth-generation fighter communications	X	X	X			X
Formation flight	X	X	X			X
Land in degraded visual environment	X		X			X
GPS-independent navigation			X			X
Aerial refueling		X				X
Minimal logistic footprint	X		X	X	X	X
Pressurized cabin		X	X			
Cargo/combat offload	X					X
PR hoist		X				X
Sensors for PR search		X				X
Real-time mission replanning	X		X			X

NOTE: GPS = global positioning system.

---

[4] The first scenario relates to the AFWERX infiltration/exfiltration and tactical mobility mission profiles, and the second scenario relates to the AFWERX AE and PR mission profiles.

Step 3. Develop a Potential Solutions List

We searched the RDT&E data set using the terms *vertical takeoff and landing* and *structure vertical lift*.[5] The top-ranked records predominantly came from five clusters that encompass five Army PEs, one Navy PE, and one joint Air Force and Navy PE (see Figure 6.1). Based on information contained in the budget documents for these PEs, we identified HSVTOL capabilities under development. The Army PEs directly address a wide range of HSVTOL capabilities, whereas the Navy and Joint PEs indirectly address a more limited set (see Table 6.2).

Figure 6.1. RDT&E Clusters Related to Vertical Takeoff and Landing

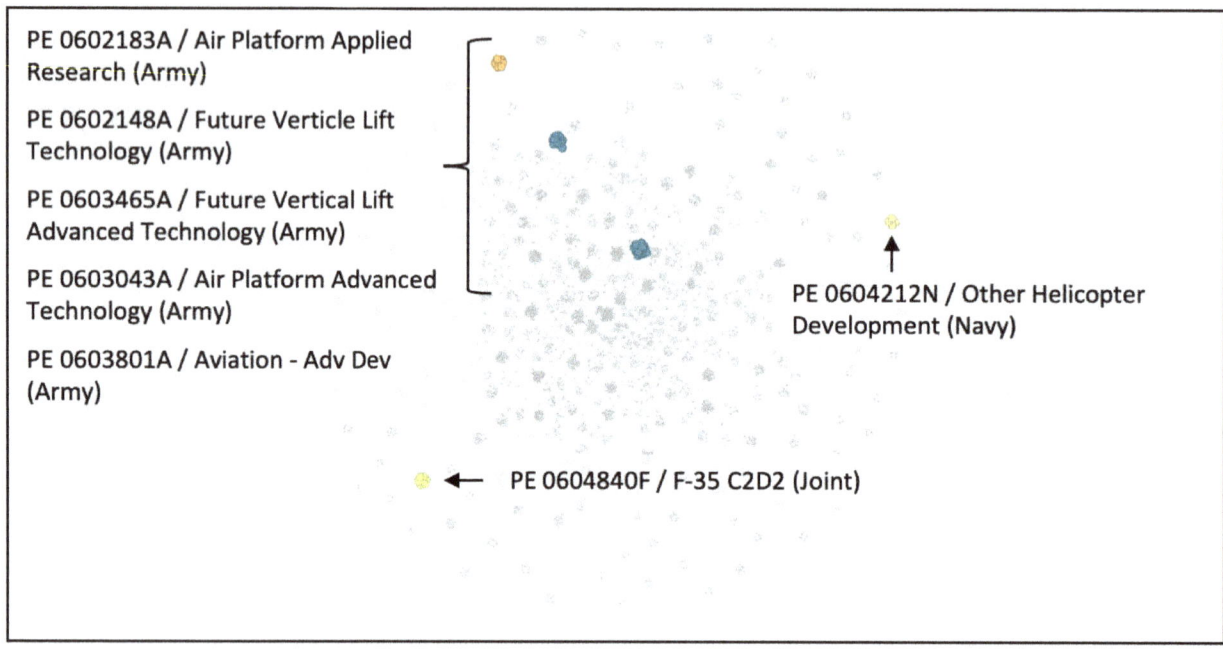

We then searched the DTIC more broadly for each of the technology capabilities listed in Table 6.2. We developed search phrases incrementally through exploration, to ensure that results returned were not overly general or narrow. Initial searches returned some unrelated records that were useful for identifying terms with multiple meanings to omit. Initial searches also returned highly relevant records that were useful for determining which words to add or retain. By drawing on domain knowledge and iterating with the data science tool in this way, search results quickly converged to highly relevant sets of records.

Except for *pressurized cabins*, DTIC contained records relevant to all the HSVTOL technologies. Figure 6.2 shows companies listed in the records along with TRLs of the technologies described. Some companies (i.e., Boeing, Raytheon, and Rockwell Collins) are developing multiple key technologies, whereas others are developing one or a small number of

---

[5] Search results were consistent using variations of these terms and phrases.

Figure 6.2. High-Speed Vertical Takeoff and Landing Technologies by Company and Maturity

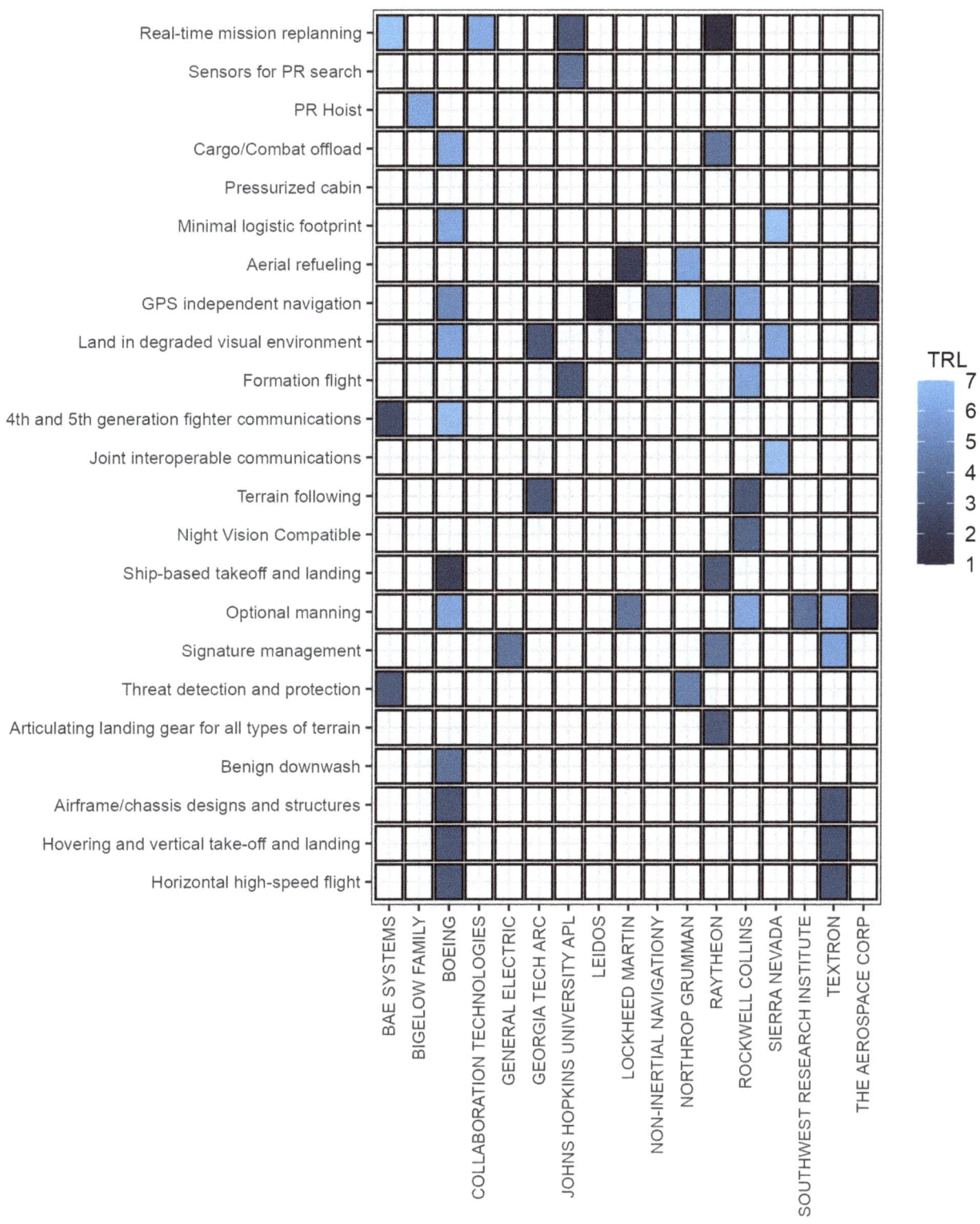

key technologies (i.e., Non-Inertial Navigation). In addition, the median TRL was lower for primary HSVTOL systems (i.e., horizontal high-speed flight, hovering and vertical takeoff and landing, and airframe/chassis design) than for secondary systems that have already been developed and applied elsewhere (i.e., joint interoperable communications, PR hoist). That said, the maturity of existing subsystems for integration on an HSVTOL platform would be expected to lag TRLs.

The results returned are not exhaustive; they are limited to records contained in the data sources we compiled. One direction for continuous development is to expand the range of data sources included to increase coverage and reduce bias in the sets of records returned by the data science tool.

Step 4. Iterate the Final Technology Solution Set

We stopped short of performing this step. Once again, one could apply an additional foresight method, such as roadmapping, to identify the remaining S&T developments needed for HSVTOL. Because the primary and secondary HSVTOL systems interact with one another in complex ways, it would also be appropriate to use a foresight method such as modeling. Finally, because the HSVTOL vignettes involve competing with an adversary it would be especially beneficial to also use red teaming.

Case Study 2. Joint All-Domain Command and Control

JADC2 seeks to link sensors, weapon systems, and networks to warfighters in all domains (land, air, sea, cyber, and space) and across all services at the scale and speed necessary for emerging joint warfighting concepts (Vergun, 2021); JADC2 is about rapidly conveying information to the right person or machine to gain information, awareness, and decision advantages over the adversary. Through heavy interconnectivity, JADC2 promises to increase resiliency in contested environments, accelerate decisions and operations at the tactical edge, integrate kinetic and nonkinetic effects, and enable long-range fires.

In this case study, we began by using SCAnTExT to retrieve formal capability gaps and informal warfighter submissions related to JADC2. Owing to the abstract nature of JADC2, the results of the initial search were not concentrated and were not practically actionable. To deal with this, we applied a foresight method—relevance trees—to decompose JADC2, a broad topic, into simpler subtopics based on recent JADC2 DoD publications. Finally, we used SCAnTExT to identify technology developments across military, commercial, and academic sectors that are relevant to the subcategories.

Step 1. Extract an Initial Capability Gap Set

We began by selecting the AFUTL as the searchable data set in SCAnTExT and entered the terms *command* and *control* to see how the Air Force databases might describe concepts associated with JADC2.[6] AFUTL records returned by the search included additional terms like *battlespace awareness*, *global information grid*, *situational awareness*, *intelligence*, *national tactical integration*, *fusion*, *network*, *decision*, and *awareness*, all of which we added to the list of search terms. We then applied combinations of these terms to the IPLs. Some combinations returned records that were clearly related to JADC2. Other terms were too general. Finally, we applied combinations of these terms to the unclassified JLLIS data. In the process, we identified several material issues with JADC2, including control authorities, applicability to agile combat employment, procedures, and training. Although these are not technology issues, they have strong implications for JADC2 systems under development.

Step 2. Iterate the Selection of Gaps

As highlighted above, the breadth of JADC2 precludes treating it as a monolithic operational capability gap. The abstract nature of JADC2 also makes it difficult to identify enabling technologies and associated R&D efforts. To gain traction, we decomposed JADC2 into the subtopics shown in Figure 6.3.[7]

Figure 6.3. Joint All-Domain Command and Control Enablers

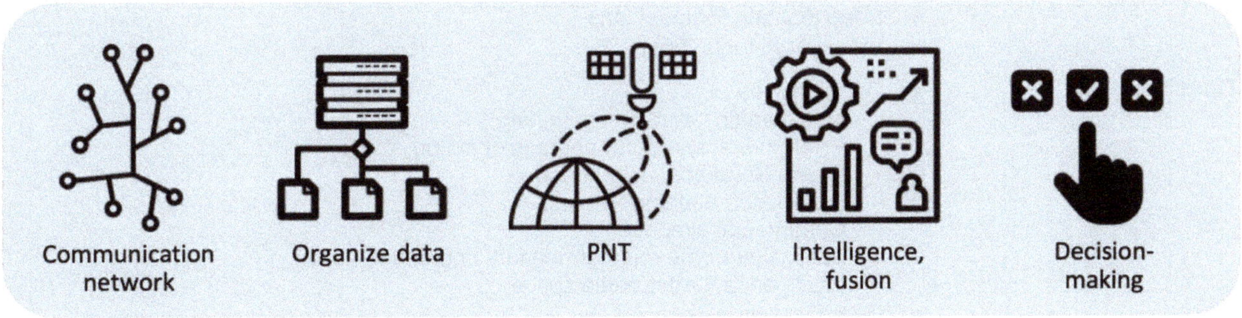

| Communication network | Organize data | PNT | Intelligence, fusion | Decision-making |

Step 3. Develop a Potential Solutions List

We used SCAnTExT to search DTIC, NSF, RDT&E, and SBIR databases for technology developments relevant to these five subtopics (see Table 6.3).[8] Some subtopics such as *organize data* and *decisionmaking* cut across academic, commercial, and governmental RDT&E. These subtopics are also general, so our searches returned diverse collections of technology enablers.

---

[6] Since the term *JADC2* is relatively new, we were unsure whether it would appear in certain data sets. However, because SCAnTExT searches based on semantic similarity, it could conceivably return records related to JADC2 even if they did not include the exact term.

[7] These partially overlap with the decomposition given by the Air Force Futures JADC2 Cross-Functional Team, and they reflect a subset of more granular technology enablers identified in conjunction with the AFRL.

[8] As in the previous case study, phrases were developed incrementally through exploration to ensure that the results returned were not overly general or narrow.

Table 6.3. Joint All-Domain Command and Control Technology Enablers

Subtopic	Technology Enablers
Communication network	• Low size, weight, and power hosted payloads for commercial satellites • Data links for manned-unmanned teaming • Inserting fifth-generation millimeter wave technology into military systems • High-speed signal processing • Troposcatter ducting • Free-space optical or high-frequency communication for contested environments
Organize data	• ML for data ingestion • Big-data solutions for data storage and management • Multilevel security • Data mining • Data fusion and big data analytics for the forward battle space.
Positioning, navigation, and timing	• Alternative navigation and positioning, navigation, and timing technologies • Antijamming antennas • Advanced waveforms and processing • Leveraging existing software-defined radios and/or commercial satellites
Intelligence, fusion	• Human-machine teaming for intelligence • Commercial AI/ML for T-sensing phenomenology and multisensor tracking • Manned-unmanned teaming • Bistatic sensing • Patterns of life in space • Deep learning for munitions • Virtual assistants for analysts

Decisionmaking	• Decision Aids
	  – Multidomain operations planning
	  – Predictive analytics for courses of action
	  – Space situational awareness
	  – Distribution and supply chain
	  – Commercial airspace
	  – Tactical airborne countermeasure courses of action
	  – Intelligence for the battle space
	• Decision-Aiding Techniques
	  – Hierarchical deep reinforcement learning
	  – NLP
	  – Knowledge graphs

Other subtopics, such as positioning, navigation, and timing, appeared more narrowly in government and government-sponsored research and contained more homogeneous technology enablers. Finally, much of the government-sponsored research was concentrated in the DAF. However, we also discovered relevant technology enablers across other services and in other federal agencies—for example, the U.S. Department of Education and the National Aeronautics and Space Administration (NASA). Although DAF research is most directly relevant, work originating from other services and federal agencies presents new opportunities to augment or go beyond DAF-sponsored research.

Step 4. Iterate the Final Technology Solution Set

Given that the concept of JADC2 is so general, the enabling technologies revealed by this case study do not address a specific operational capability gap. Consequently, foresight methods intended for open-ended problems would be especially applicable in this final step. For example, science fiction may reveal innovative applications of these technologies to JADC2, while wild cards/weak signals may allow the services to define low-probability but high-impact technology breakthroughs that would transform future JADC2 operations.

Case Study 3. Human Capital Management

The goal of Air Force human capital management (HCM) is to ensure the readiness of airmen to fulfill the Air Force mission. Air Force HCM efforts include recruiting, selection, job classification, career development and education, assignment, promotion, and retention. Chief of Staff of the Air Force Charles Q. Brown Jr. recently articulated the importance of HCM, and specifically the importance of adopting modern approaches to it, stating that the Air Force must leverage "modern information technology approaches to enhance and deliver talent management solutions," and "to make analytically informed and timely decisions" (Brown, 2020).

In this case study, we began by using SCAnTExT to retrieve formal descriptions of Air Force HCM needs along with informal warfighter submissions about HCM limitations. We settled on the capability gap and general technology solution articulated by Chief of Staff of the Air Force Brown: to use modern IT approaches to improve HCM. We then combined a pair of foresight methods—scenario creation and brainstorming—to produce a more complete and actionable description of the capability gap. Scenario creation entails generating a "story" to illustrate a vision of a possible future. Brainstorming involves freely generating ideas and then combining and rank ordering them. Finally, we used SCAnTExT to identify technology developments across military, commercial, and academic sectors relevant to the future end state.

Step 1. Extract an Initial Capability Gap Set

We began by setting the AFUTL as the searchable data set in SCAnTExT; and we entered the words *personnel*, *training*, and *education*.[9] Two of the top results were: "Provide personnel and education, and training program support" and "Generate personnel." We repeated the search using Air Mobility Command's Office of the Surgeon General and PACAF CBA data sets. Once again, the top results in both data sets overwhelmingly related to force generation and management—for example, "Increase career development as operational/clinical personnel" and "Determine best mechanism to recruit, train and retain required personnel to meet AE mobility

---

[9] Search results were robust across variations of these and related terms.

planning requirements." The MAJCOM and CCMD IPLs tended to focus on operations rather than mission support and so did not include relevant entries.

Next, we turned to JLLIS for crowdsourced HCM capability limitations. When we used the same search terms, SCAnTExT primarily returned results related to training, reflecting the large number of such entries that JLLIS contains.[10] To narrow the results, we set the reference data set to the AFUTL and searched JLLIS based on text from the two most relevant AFUTL records: "Generate forces" and "Provide personnel." Of the top 20 results, six were about whether individuals had completed experiences that prepared them for assignments, seven were about creating occupation-specific developmental pathways, and six were about tracking work histories of civilian personnel.[11] For example, one JLLIS record stated,

> Introducing personnel to the AOC [Air Operations Center] weapon system at the command or senior leadership level and expecting them to do well isn't working. . . . A high percentage of AOC CCs and Division leaders reported they had never been in an AOC prior to assuming their command or leadership position. In most weapon systems, the USAF "grows" leaders through multiple and relevant career broadening assignments.

To summarize, SCAnTExT returned formal descriptions of capability gaps related to force generation and management and informal warfighter submissions related to HCM. These records contextualize and enrich the capability gap articulated by Chief of Staff of the Air Force Brown, and they can be used for a variety of purposes including scenario creation.

Step 2. Iterate the Selection of Gaps

We used foresight methods to build on the initial capability gap list—which, in this case, was limited to the use of modern IT approaches to improve HCM. Based on the information returned by SCAnTExT, combined with SME knowledge, we generated a scenario focused on officer assignments in the time frame set forth in the S&T 2030 strategy document (DAF, 2019):

> The airman was eligible to move during the next assignment cycle. She considered her career progression and settled on the importance of gaining operational level C2 experience. She logged into the Talent Marketplace web-based system and updated her knowledge, skills, ability, and other characteristics. She then entered preferences based on location and position for her next assignment, prioritizing AOCs.
>
> After the airman and all others who were eligible for assignment placed bids, commanders logged into the system to view who volunteered to fill vacancies. Each commander rank ordered the list of interested officers. Upon reviewing the airman's materials, an AOC commander recognized her potential and ranked her highest for a position as an AOC division director.

---

[10] Of the first 100 results returned, all but one involved training.

[11] JLLIS contained records related to other HCM processes. Given how differently the processes are described, different phrases were needed to locate those records.

Next, the Air Force Personnel Center ran a pairing algorithm that took candidate and commander preferences into account. In addition, the algorithm considered the candidate's complete assignment history, narrative data from annual performance reports, narrative data from the losing commander, and other variables from administrative databases. To do so, the algorithm securely accessed data contained in the Talent Marketplace and from across the enterprise. From these data, the algorithm predicted each candidate's performance and growth potential for every possible assignment. The algorithm predicted that the airman would perform well as an AOC division director given her experience, and that the assignment would contribute to her career growth. The algorithm performed a constrained optimization to maximize preference matching, job performance and career growth, and recommended the airman for the AOC assignment.

Finally, career field managers reviewed the algorithm's recommendations and finalized assignment decisions. To increase trust and improve the performance of the joint human-machine team, the algorithm provided information to explain its recommendations to career field managers. The career field managers accepted the recommendation, and the airman was assigned to the position as an AOC division director.

A group of SMEs then determined a partial set of the technologies needed to arrive at the future state described in the scenario (i.e., brainstorming). The SMEs identified seven key technologies, divided among three groups:

1. Algorithms
   a. *NLP* to extract information from narrative data
   b. *ML* to create models to predict job performance
   c. *optimization techniques* to maximize preference matches, job performance, and career growth
2. Interfaces
   a. *explainable AI* to allow human decisionmakers to understand recommendations
   b. *user interfaces* to allow human decisionmakers to enter additional decision information and explore different assignment courses of action
3. Infrastructure
   a. *secure infrastructure* for data housing and information processing
   b. *secure developmental pipeline* for automating software development and ML workflow.

Step 3. Develop a Potential Solutions List

Based on the outputs from Steps 1 and 2, we used SCAnTExT to find relevant technology developments across military, commercial, and academic sectors.[12] We focused on three of the key

---

[12] As in the previous case studies, phrases were developed incrementally through exploration to ensure that results returned were not overly general or narrow.

technologies: human-machine interfaces, NLP, and a secure software development pipeline.[13] We searched four databases for records related to these technologies: the AFRL, AFVentures, the NSF, and the SBIR. Figure 6.4 shows the number of matches per topic and source.

Figure 6.4. The Number of Records from Each Source Related to Key Technologies

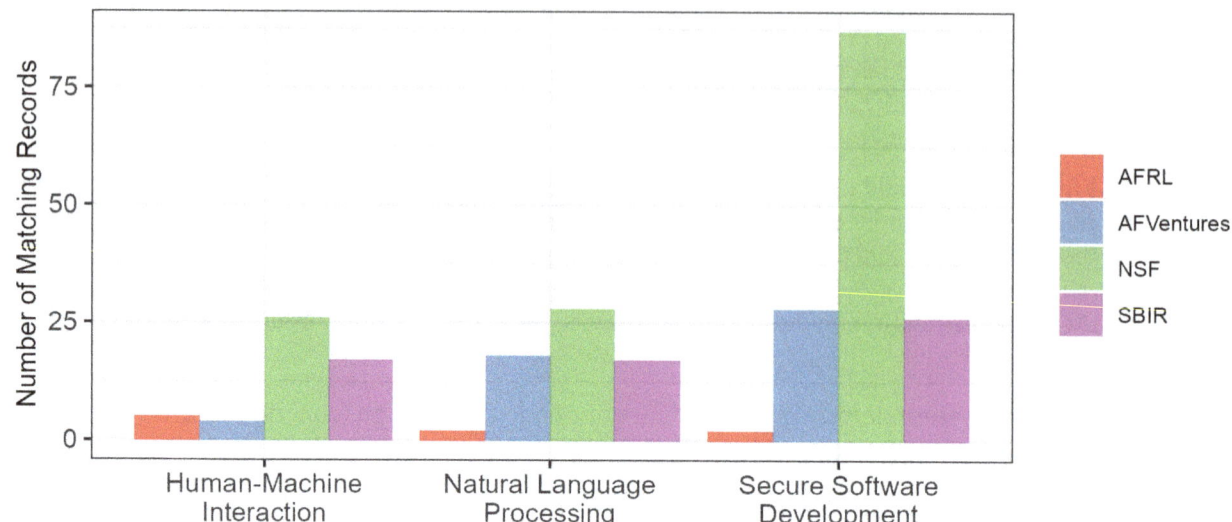

The search results support several findings:

- Work on all three technologies is being performed in government (AFRL), commercial (AFVentures and SBIR), and academic sectors (NSF and SBIR). Of the three technologies, the greatest number of matches were for secure software development, followed by NLP and then by human-machine interaction.
- The greatest number of matches for each technology are from the NSF data set, which also contains the most records overall. The SBIR data set contains a similar number of records overall yet had fewer matches for each topic. The AFRL and AFVentures data sets contain far fewer records overall, yet both contain multiple matches per topic.
- The content of records varies by source. AFVentures records describe a company's offerings (e.g., NLP), SBIR records describe a product (e.g., an NLP tool kit to parse unstructured manufacturer data), and NSF records describe a scientific approach (e.g., deep neural networks to represent language structure). AFRL records reflected a mix of scientific approaches and warfighter applications.
- DAF-sponsored work is most directly relevant. Research originating from other sources varies in relevance, but it presents untapped opportunities to augment and go beyond DAF-sponsored work.

---

[13] For NLP, we used the key words *natural language processing (NLP) natural language understanding (NLU)*. For human-machine interfaces, we used the key words *human-machine teaming man-machine teaming adaptive interface explainable AI*. For secure software development, we used the key words *secure software development devops devsecops*. An analyst reviewed the rank-ordered search results, read the abstracts, and exported relevant entries.

- The SBIR program is structured in phases: the objective of Phase I is to establish technical merit and feasibility, and the objective of Phase II is to continue the R&D effort. Forty-four percent of matches for secure software development were in Phase II, versus only 33 percent and 18 percent, respectively, for human-machine interfaces and NLP. This may reflect the relative maturity of the latter two technology areas.
- The searches led to adjacent topics that, although not identified by SMEs, may constitute additional key technologies. For example, the search for secure software development also led to clusters of results on secure ML, secure scientific computing, and data security; the search for human-machine interaction led to a cluster of results on algorithmic bias. Although these clusters only partially overlap with our search phrases, they are semantically related and practically relevant.

Step 4. Iterate the Final Technology Solution Set

We stopped short of performing this step. Using the technology areas identified in Step 3, one could apply an additional foresight method such as roadmapping to identify the remaining S&T developments needed for data-enabled HCM. In addition, HCM is strongly influenced by policy and culture. For this reason, it would also be useful to use a more wide-ranging foresight method, for example, backcasting, to identify the complete set of doctrine, organization, training, materiel, leadership, personnel, facilities, and policy steps needed to arrive at the desired future state.

Chapter Summary

Defense technology innovation—the implementation of new technologies, or of existing technologies in new ways, to meet current and future warfighter needs—depends on human expertise and creativity. Thus, capability development calls for human-centered planning and decisionmaking. Human knowledge factored into the three case studies in many ways: inputs to the initial capability gap sets originated from humans and reflected implicit prioritizations; through creative foresight methods, humans elaborated on capability gaps to define a desired future state; through analytic foresight methods, humans decomposed the future state into actionable S&T capability requirements; and through expertise-based foresight methods, humans defined technology roadmaps and additional doctrine, organization, training, materiel, leadership and education, personnel, facilities, and policy steps to arrive at the future state.

Notwithstanding, here, the central role of humans, the volume of potentially relevant information contained in text sources is too great for human processing alone. Thus, capability development also calls for data-enhanced planning and decisionmaking. Data science capabilities factored into the three case studies in many ways: formal sources and informal warfighter submissions provided support for the capability gaps that we explored; these sources also provided additional contextual details and enriched the descriptions of the capability gaps; and S&T data sources provided information about the maturity of enabling technologies, as well as the sectors and organizations leading their development.

A final, and somewhat unexpected, outcome from the case studies is the extent to which human and machine intelligence are complementary. SCAnTExT occasionally returns results that clearly differ from the user's intent (e.g., a search for *joint all domain* returned NSF awards granted to orthopedic surgeons). The user can easily identify these errant records and refine the search terms. On the other hand, SCAnTExT gives equal attention to all records, and sometimes returns unexpected but relevant results (e.g., a search for *software development pipelines* also returned clusters of results on ML development pipelines and secure databases). Across the case studies, the human and machine interaction refined how the databases were searched, while at the same time refining how the operational capability gaps and technology solutions were understood.

An underemphasized point in our analyses is that capability gaps and technology solutions originate from different organizational contributors, and they align with different organizational roles (e.g., AFRL core technical competencies or Air Force Futures Cross-Functional Teams). Organization is included as a meta tag in many of the data sources, and thus could be used to identify SMEs and stakeholders and to shape development, acquisition, and adoption strategies.

The outcomes from these case studies point to the utility of an HCDE process. Yet our evaluation of the data-enhanced capability emphasized the number and diversity of records returned. Records vary in terms of quality and relevance due to limitations in the data sources and limitations in existing NLP methods. Future work is needed to evaluate the effectiveness of SCAnTExT—and HCDE decisionmaking more generally—for improving development planning. A natural next step would be to repeat each case study, along with more operational and S&T SMEs, and to carry the HCDE process through all four steps.

Chapter 7. Findings and Recommendations

In this project we have examined how the TCO could organize and improve the way it decides which concepts to feed into the transformational capability pipeline. Our research supports seven primary findings.

1. **The TCO has an exceptionally broad mandate.** The TCO must track operational capability gaps, track S&T solutions, and prioritize development efforts. Additionally, the scope of capability gaps and S&T solutions are not limited to one functional area or technical discipline. For comparison, most other organizations we spoke with are primarily concerned with finding S&T solutions for a constrained set of capability gaps (e.g., intelligence, surveillance, and reconnaissance), or with finding capability gaps that can be addressed by a constrained set of S&T solutions (e.g., AI/ML). For this reason, the TCO may need a more structured decisionmaking process using different tools and methods than other S&T organizations. In addition to facilitating internal processes, these tools and methods may allow the TCO to better leverage products and expertise in the broader AFRL and DAF ecosystem.

2. **Certain data sources for capability gaps are widely used, but they are not centrally managed.** CGLs and IPLs are widely used for identifying capability gaps. However, the DAF does not maintain a single repository for these sources, and it does not standardize how they are reported. Warfighter submissions to crowdsourced systems like JLLIS also exist but are not widely used by S&T organizations.

3. **No software tools are used to parse, extract, and summarize the content of capability gap sources.** Formal capability gaps number in the hundreds, and crowdsourced warfighter submissions number in the tens of thousands. No software tools are systematically used to organize and extract information from these sources.

4. **Data sources for S&T solutions are far more numerous and diverse than for capability gaps.** Data sources for identifying S&T solutions originate from within the DoD, industry, and academia. Some DoD organizations use spreadsheets or proprietary databases to track S&T programs, but the DoD does not maintain a single repository of S&T projects, and it does not standardize how they are reported. The challenge of navigating the S&T space is exacerbated by the fact that so much relevant work occurs in academia and industry.

5. **Nascent tools for bibliometrics, patent analysis, landscape analysis, and horizon scanning exist but are not yet widely used.** As with capability gaps, no software tools are systematically used to organize and extract information from S&T sources.

6. **Development planning is a human-centered endeavor.** Today's S&T organizations rely on SMEs to identify capability gaps and S&T programs, and they synthesize S&T solutions. In a sense, this is necessary given the incomplete and variable-quality information contained in data sources and the background knowledge and creativity that SMEs must draw upon to augment them. Development planning is also a social endeavor. Social networks bridge the gap between individuals with deep operational knowledge and those with deep S&T knowledge. The interplay between these communities is essential to ensure that solutions are operationally and technically sound, but also to establish

widespread buy-in. For these reasons, human decisionmakers must remain part of development planning.

7. **Informal methods are used to identify capability gaps and S&T programs, and to synthesize solutions.** The ways in which SMEs identify capability gaps and S&T programs and synthesize S&T solutions are often informal and unstructured (e.g., through interpersonal communication or brainstorming). Some organizations we spoke with expressed concern that the methods used are laborious, incomplete, and prone to bias. A collection of techniques called foresight methods exist for making future-focused decisions. Relatively few organizations we spoke with reported using these methods, and those that did used only a few common methods. Wider application of a richer set of these foresight methods to the various phases of development planning may therefore increase the utility of human outputs.

These findings motivate the flow diagram shown earlier in Figure 3.1, which represents the vertical integration among HCDE decision processes and the lateral integration between identifying capability gaps and technology solutions. This is a workable model for applying the data sources, the data science tools (i.e., SCAnTExT), and the foresight methods considered in this report.

A final finding is that the challenge of bridging the "valley of death" remains. While this goes beyond the scope of our report, we note that most organizations we spoke with felt that solving the technology transfer problem was at least as important as selecting the right transformational concepts to begin with.

Recommendations

Based on these findings, we offer several recommendations. The primary recommendation is to use an HCDE process for development planning, and the subsequent recommendations involve steps to reach that end state.

1. **The AFRL and the TCO should use the concept development and selection process illustrated in Figure 6.1, or a variant thereof.** Specifically, information from natural language descriptions of capability gaps and S&T solutions should be integrated into a human-centered decision process. This is the primary recommendation.

2. **The AFRL and the TCO should use a tool like SCAnTExT to allow human decisionmakers to extract information from natural language data sources.** Such a tool can reveal new and emerging clusters of capability gaps and technology solutions not yet being considered, and it can provide additional information about those that are. Such a tool can also identify the individuals, companies, organizations, and other stakeholders (i.e., a social network) engaging with a capability gap or S&T solutions. In addition, as the AFRL and the TCO use data science methods to extract information from natural language data sources, they should conduct usability and validation studies to evaluate and advance software systems.

3. **The AFRL should explore alternate NLP methods and implementations for using free-text data.** The core NLP method in SCAnTExT, LSA, is well established and has been shown to be competitive with more recent techniques for text mining from security-

and defense-related sources (Schirmer, 2021). Even so, different parameterizations of LSA and alternate NLP methods may be more suitable for extracting information from operational capability gap and S&T solution data sources. The AFRL should evaluate a broader set of approaches to determine which produce the most valuable insights.

4. **The DAF should curate and standardize key data sources.** There is no single source/repository for DAF operational and technical capability gaps. A central repository of key documents explicitly stating capability gaps should be established. The repository should include existing command needs (e.g., CGLs and IPLs) and functional needs (e.g., the ISR Capabilities Analysis Requirements Tool). In addition, capability gaps should be reported in a standardized manner across commands and functional communities to facilitate analysis. Finally, the repository should be augmented with statements of future-focused needs originating from Air Force Futures.

5. **The AFRL, the DAF, and the TCO should enrich key data sources.** The descriptions of S&T programs contained in budget documents and POM submissions are a valuable source of information, yet the quality of the descriptions varies widely. The AFRL and the DAF should incentivize submitters to ensure that descriptions are accurate, current, and complete. In addition, S&T data sources from industry and academia are incomplete and contain errors. The AFRL and the TCO should purchase cleansed sources and metadata that are commercially available or invest in software to cleanse sources and generate metadata internally. Finally, the DAF uses specialized language with many semantically interchangeable or related words and terms (e.g., *JADC2* and *Joint All-Domain Command and Control*). The AFRL should create a lexical database (i.e., a WordNet) to standardize terminology contained in text records during data processing.

6. **The TCO should expand the use of foresight methods.** Two methods, brainstorming and science fiction, were found to be applicable to all TCO initiatives. Of these, only brainstorming is widely used today. The TCO should use science fiction along with other creative, interactive, expert-driven, and evidence-based methods tailored to different programs to generate transformational concepts. Other offices within the AFRL (e.g., technical directorates or the Air Force Office of Scientific Research) or within the DAF should use the method described in Chapter 5 to select foresight methods for programs focused on different types of capabilities.

7. **As a stepping stone to reach full curation and standardization of HCDE capability development planning, the AFRL and the TCO should record human-generated technology pairings for capability gaps.** A limitation of the data science tool is that it provides shallow mappings from capability gaps to S&T solutions. By recording human-generated solutions for capability gaps and embedding them in the data science tool, the tool may recommend technology sets for new problems based on analogy to existing problems. The AFRL and the TCO should consider converting program outputs, proposals, and other planning outputs into a standardized machine-interpretable form to allow the data science tool to provide deep mappings between future capability gaps and S&T solutions.

Conclusion

Although the data science tool (SCAnTExT) and foresight methods described here were developed and selected for the TCO, all elements of the methodology can be used by other organizations. The data science tool is broadly suitable for searching through text data and grouping records, whether they originated from the DAF or elsewhere. In addition, the approach we used for selecting foresight methods for different TCO initiatives can be repeated for non-TCO initiatives as well. Figure 7.1 shows the more general application of these principles to HCDE decisionmaking.

Figure 7.1. The General Approach for Human-Centered, Data-Enhanced Decisionmaking

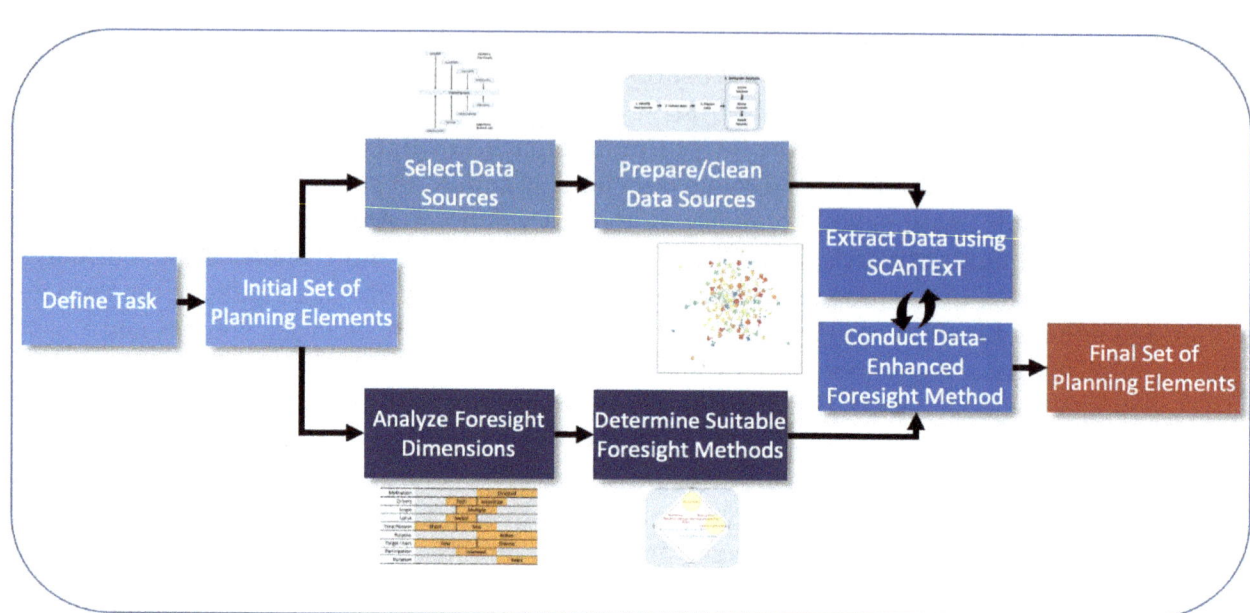

The general approach begins by defining the task—whether that is to identify capability gaps, S&T solutions, or something else related to concept development. Next, a starting set of planning elements are provided to determine which data sets may be useful and to establish dimensions of the foresight problem. Data sources are then selected and cleansed using NLP techniques. In parallel, foresight methods are selected based on the dimensions of the foresight problem. Finally, a tool like SCAnTExT is used to extract information from data sources and to enable data-enhanced foresight methods. The final output of this HCDE decision process is a collection of recommendations for RDT&E actions—for example, which concepts to pursue in a portfolio of investments. We expect that this general design for HCDE decisionmaking is applicable to other organizations besides the TCO that also seek innovative technology solutions for the most significant warfighter challenges.

Appendix A. Interview Protocol

Background

[I would like to start by asking you a few questions about your background]

A.1. I see that you work in [office]. Is that correct?

    1. For how long and in what role?

A.2. Can you please describe your office's mission with respect to identifying warfighter needs, developing technology solutions, and/or applying technology solutions?

A.3. With that in mind, what would you say are your office's percentages of effort across the following three activities?

    1. Identifying warfighter capability gaps
    2. Developing technologies
    3. Applying technologies to warfighter capability gaps

A.4. What time frame does your office consider when identifying warfighter needs or technology solutions?

A.5. Is there a particular type of capability gap or technology that your office primarily considers?

Organizational Process Questions

[For the next set of questions, I would like to ask you about warfighter capability gaps]

B.1.1. What data sources and processes do you use to identify warfighter capability gaps?
    Probes (asked as needed):

    1. What are the strengths of these data sources and processes?
    2. What are the limitations of these data sources and processes?

B.1.2. What information do you collect about warfighter capability gaps?

B.1.3. What tools do you use to manage the data and to conduct these processes?

B.1.4. What criteria do you use to prioritize warfighter capability gaps?
    Probe (asked as needed):

    1. How do you evaluate these criteria?

*[**For the next set of questions, I would like to ask you about identifying technology programs that could address those gaps**]*

B.2.1. What data sources and processes do you use to identify technology programs?
Probes (asked as needed):

1. What are the strengths of these data sources and processes?
2. What are the limitations of these data sources and processes?

B.2.2. What information do you collect about technology programs?

B.2.3. What tools do you use to manage the data and to conduct these processes?

B.2.4. What criteria do you use to prioritize technology programs?
Probe (asked as needed):

1. How do you evaluate these criteria?

*[**For the next set of questions, I would like to ask you about prioritizing technology solutions**]*

B.3.1. How do you map technology programs to capability gaps?

B.3.2. What criteria do you use when prioritizing technology solutions?
Probe (asked as needed):

1. How do you evaluate these criteria?

Closing Questions

*[**I would like to end by asking you two closing questions**]*

C.1. If you could change anything about the Air Force's approach to capability development planning, what would it be?

C.2. Can you describe a success story from within your office?

Appendix B. Foresight Methods

Descriptions of Foresight Methods

Table B.1 contains descriptions of 33 foresight methods divided among three categories: qualitative, quantitative, and semiqualitative.

Table B.1. Descriptions of Foresight Methods

Type	Method	Description
Qualitative	Backcasting	Working backward from an imagined future to determine the path to that future from the present
	Brainstorming	Using group working sessions to generate new ideas around a specific area of interest
	Citizen panels	Gathering views on relevant issues from groups of citizens
	Conferences	Events and meetings that typically involve a mix of talks and discussions on a particular subject
	Essays	Producing plausible accounts of future events and developments based on systematic analysis of the present
	Expert panels	Bringing together people to combine knowledge in their areas of shared expertise
	Genius forecasting	Combining creativity and analysis to prepare a future forecast in one's area of expertise
	Interviews	Structured conversations used to gather knowledge that is distributed across the range of interviewees
	Literature review	Discursive review to summarize work on a particular subject
	Morphological analysis	Mapping promising solutions to a given problem and determining the possible futures that may result
	Relevance trees	Subdividing a large subject into smaller subtopics to produce a visual hierarchical structure of the subject
	Role-play	Participants imagine and act as though they were other states or individuals
	Scanning	Observing, monitoring, and systematically describing technological, sociocultural, political, or ecological changes
	Scenarios	Combining several features or variables to construct various scenarios
	Science fiction	Generating stories about possible events that may materialize at some point in the future
	Simulation gaming	Role-play based on a script that outlines the context, actions, and actors involved
	Surveys	Using questionnaires to gather knowledge distributed across a range of respondents

Type	Method	Description
	SWOT	Identifying and classifying factors internal to an organization into strengths and weaknesses, and then identifying and classifying factors external to an organization into opportunities and threats
	Wild cards/weak signals	Identifying future high-impact situations or events with a low probability of occurring (i.e., wild cards), and defining observable warnings about the probabilities of those events (i.e., weak signals)
Quantitative	Benchmarking	Comparing what other organizations are doing in comparison with what one's own organization is doing
	Bibliometrics	Quantitative analysis of publications
	Modeling	Using computer-based models to simulate outcomes produced by different combinations of state or system variables
	Patent analysis	Quantitative analysis of patents
	Time series analysis	Analyzing indicators to measure change over time
	Trend extrapolation	Forecasting future developments based on assumptions about the continuation of past and present developments
Semiquantitative	Cross-impact/structural analysis	Systematic evaluation of combinations of variables, commonly represented in a matrix, that determine the behavior of a system
	Delphi method	Repeated polling of the same individuals along with anonymized exchange of responses to allow for consensus building
	Key technologies	Listing technologies that will contribute to the success of an organization, sector, country, or region
	Multicriteria analysis	Asking participants to assess the importance of multiple evaluative criteria for a complex problem and to assess the impact of different options or strategies on those criteria
	Polling/voting	Survey method to assess the strength of views about a particular topic among a set of respondents
	Quantitative scenarios	Quantification of contingencies that bring about a scenario, and evaluation of the likelihood of occurrences of certain events
	Roadmapping	Creating detailed projections of possible technology developments, products, or environments to arrive at a desired future state
	Stakeholder analysis	Identifying key objectives of different stakeholders and representing them in the form of matrixes

SOURCE: Popper, 2008.

Dimensions of Foresight Problems and Mapping to Methods

Porter (2010) identified nine dimensions to characterize foresight methods and endeavors. Each dimension has multiple state values. Tables B.2 and B.3 show the state values that each foresight method is suitable for.

Considerations for Use of Foresight Methods Along Motivation, Driver, Scope, Locus, and Time Horizon Dimensions

	Motivation	Drivers	Scope	Locus	Time Horizon
	Directed	Science, technology, innovation, context	Single, multiple, wide-ranging	Institution, sector, national	Short, medium
۱g	Open-ended, directed	Science, technology, innovation, context	Single, multiple, wide-ranging	Institution, sector, national, global	Short, medium, long
۹ls	Directed	Technology, context	Single, multiple	Institution, sector, national	Short, medium
s	Open-ended, directed	Science, technology, context	Single, multiple, wide-ranging	Institution, sector, national, global	Short, medium
	Open-ended	Science, technology, context	Single, multiple, wide-ranging	Institution, sector, national, global	Short, medium
ls	Open-ended, directed	Science, technology, context	Single, multiple	Institution, sector, national, global	Short, medium
casting	Open-ended	Science, technology, innovation, context	Single	Institution, sector, national	Short, medium
	Open-ended, directed	Science, technology, context	Single, multiple	Institution, sector, national, global	Short, medium
·view	Open-ended, directed	Science, technology	Single, multiple	Institution, sector, national, global	Short, medium
ـal analysis	Directed	Technology, innovation	Single, multiple	Institution, sector, national, global	Short, medium
·rees	Directed	Technology, context	Single, multiple, wide-ranging	Institution, sector, national, global	Short, medium
	Open-ended	Technology, context	Single, multiple, wide-ranging	Institution, sector, national, global	Short, medium
	Open-ended, directed	Science, technology, context	Single, multiple	Institution, sector, national, global	Medium, long
	Open-ended, directed	Technology, context	Single, multiple, wide-ranging	Institution, sector	Short, medium, long
·tion	Open-ended, directed	Science, technology, innovation, context	Single, multiple, wide-ranging	Institution, sector, national, global	Medium, long
۱aming	Directed	Technology, context	Single, multiple	Institution, sector	Short, medium

od	Motivation	Drivers	Scope	Locus	Time Horizon
eys	Open-ended, directed	Science, technology, context	Single, multiple, wide-ranging	Institution, sector, national, global	Short, medium
T	Open-ended, directed	Technology, context	Single, multiple	Institution, sector, national	Short, medium
cards/weak ls	Open-ended	Science, technology, innovation, context	Single, multiple	Institution, sector, national, global	Medium, long
hmarking	Directed	Technology	Single, multiple	Institution, sector, national, global	Short, medium
metrics	Open-ended, directed	Science	Single, multiple	Institution, sector, national, global	Short, medium
eling	Directed	Science, technology, context	Single	Institution, sector	Short, medium
t analysis	Open-ended, directed	Technology	Single, multiple	Institution, sector, national, global	Short, medium
series analysis	Open-ended, directed	Science, technology, context	Single, multiple	Institution, sector, national, global	Short, medium
extrapolation	Open-ended	Science, technology, context	Single, multiple	Institution, sector, national, global	Short, medium
s-impact/structural sis	Directed	Technology, context	Single, multiple	Institution, sector, national, global	Short, medium
i method	Open-ended, directed	Science, technology, context	Multiple, wide-ranging	Institution, sector, national, global	Medium, long
echnologies	Directed, Open-ended	Technology	Multiple	Institution, sector, national	Short, medium
criteria analysis	Directed	Technology, context	Multiple, wide-ranging	Institution, sector, national, global	Short, medium
g/voting	Directed	Science, technology, context	Single, multiple	Institution, sector	Short, medium
titative scenarios	Directed	Technology, context	Single, multiple	Institution, sector	Short, medium
mapping	Directed	Technology	Single, multiple	Institution, sector, national, global	Medium
eholder analysis	Directed	Technology, innovation, context	Multiple, wide-ranging	Institution, sector, national, global	Short, medium

.3. Considerations for Use of Foresight Methods Along Purpose, Target Users, Participation, and Study Duration Dimensio

	Purpose	Target Users	Participation	Study Duration
sting	Information	Few, diverse	Narrow, intermediate	Months, years
orming	Information	Few, diverse	Narrow, intermediate, diverse	Days, months, year
panels	Information	Diverse	Intermediate, diverse	Months, years
nces	Information	Few, diverse	Intermediate, diverse	Months, years
	Information	Few, diverse	Narrow, intermediate	Days, months, year
panels	Information	Few, diverse	Narrow, intermediate, diverse	Months, years
forecasting	Information	Few, diverse	Narrow	Days, months, year
ws	Information	Few, diverse	Intermediate, diverse	Days, months, year
re review	Information	Few	Narrow, intermediate	Days, months, year
logical analysis	Information, action	Few	Intermediate	Months, years
ce trees	Information, action	Few	Intermediate	Months, years
ay	Information	Few, diverse	Intermediate, diverse	Months, years
ng	Information	Few, diverse	Narrow, intermediate, diverse	Months, years
os	Information	Few, diverse	Intermediate, diverse	Months, years
Fiction	Information	Few, diverse	Narrow, intermediate, diverse	Days, months, year
ion gaming	Information	Few, diverse	Intermediate, diverse	Months, years
s	Information	Few, diverse	Intermediate, diverse	Months, years
	Information, action	Few, diverse	Intermediate	Months, years
rds/weak signals	Information	Few	Narrow, intermediate	Months, years
narking	Action	Few	Narrow	Months, years
etrics	Information	Few	Narrow	Months, years
g	Information, action	Few	Narrow, intermediate	Months, years
analysis	Information	Few	Narrow	Months, years
ries analysis	Information	Few	Narrow, intermediate	Months, years
xtrapolation	Information	Few	Narrow	Months, years

71

Method	Purpose	Target Users	Participation	Study Dura
Cross-impact/structural analysis	Action	Few, diverse	Narrow, intermediate	Months, year
Delphi method	Information	Few, diverse	Intermediate, diverse	Months, year
Key technologies	Information	Few, diverse	Intermediate	Months, year
Multicriteria analysis	Action	Few, diverse	Narrow, intermediate	Months, year
Polling/voting	Information	Few, diverse	Intermediate, diverse	Months, year
Quantitative scenarios	Information, action	Few	Narrow, intermediate	Months, year
Roadmapping	Information, action	Few	Narrow, intermediate	Months, year
Stakeholder analysis	Information, action	Few, diverse	Narrow	Months, year

Contemporary Trends in Foresight Across Sectors

Horizon Scanning and Scenario Planning

Foresight methods are commonly used in the federal government. For example, a recent study surveyed federal foresight efforts and ways to better integrate foresight methods into strategic planning and decisionmaking (Greenblott et al., 2019). In this study, the authors examined the results of semistructured interviews with individuals involved with foresight in 19 federal agencies, as well as two nonfederal experts on foresight in government.

The most frequently used foresight methods across federal agencies included horizon scanning and scenario planning. Horizon scanning involves a "systematic process for gathering and analyzing information on trends and emerging or potential developments that may be important for an organization, including new threats, additional responsibilities, or untapped opportunities" (Greenblott et al., 2019, p. 250). Many organizations used horizon scanning to summarize findings for leadership, as well as to inform scenario planning. Scenario planning pushes organizations to think of alternative descriptions of how the future may unfold, including how they would respond to such conditions and situations. Other foresight methods mentioned during our interviews included speaker series, the Delphi method, formal foresight training, backcasting, and future wheels.

Foresight methods are also commonly used in the commercial sector. For example, one study surveyed 14 foresight practitioners who represented American corporations or American divisions of European corporations. The most frequently used foresight methods across the 14 corporations included scenario planning, trend analysis, environmental scanning, workshops, and wild cards/ weak signals (Hammoud and Nash, 2014). Scenario planning was often used to identify key topics, discuss strategies based on various scenarios, and learn about signals of change for research purposes. As for trend analysis, corporations used this form of foresight to understand developments in macroenvironments over the course of two to three years, investigate clusters of smaller trends with significance, and establish big trends. Environmental scanning was used to identify signals of change, new developments, important trends, and any major points of discontinuity on the horizon. With respect to workshops, respondents claimed that these forums helped facilitate ideation and bolster communication with the corporation. As for wild cards/ weak signals, respondents shared how the method was used by scanning blogs, news groups, external research companies, and daily newsletters to reveal new perspectives, weaknesses, and to challenge existing assumptions. Moreover, workshops allow individuals within corporations to have dialogues about innovation and the varied needs of different consumer markets.

Finally, foresight methods are commonly used in different countries. For example, one study of European foresight efforts described over 30 methods (Popper, 2008). The study underscored how selecting methods is a multifactor process that draws on the fundamental attributes (e.g., creativity, evidence, interaction, and expertise) and nature (e.g., quantitative, qualitative, and semiquantitative) of different methods. The study also emphasized the concept of a "methods mix," which refers to the combination of foresight methods used to tackle foresight problems. For example, the average number of methods used per project in European government, academic, and business contexts ranged from four to five.

The most widely used methods in Europe included literature reviews, expert panels, and scenarios. These all draw on qualitative approaches. Other commonly used methods included extrapolation/megatrends, future workshops, brainstorming, interviews, the Delphi method, questionnaires/surveys, key technologies, scanning, essays, and SWOT. Finally, the less frequently used methods included roadmapping, modeling/simulation, backcasting, stakeholders mapping, structural analysis, bibliometrics, morphological analysis, citizens panels, relevance trees, multicriteria analysis, and gaming.

Appendix C. Data Science Methods

In this appendix, we describe how the natural language data sources were preprocessed. We then describe two extensions to the NLP path. The first involves manually labeling a subset of records and applying supervised learning techniques to those data. The second involves using pretrained language models, an active area of NLP research, rather than LSA to extract information from records.

Data Preprocessing

All NLP analyses were performed in the R computing language using the textmineR package (R Foundation, 2020). We followed well-established steps for preparing text data for ML. First, we divided each record (i.e., a text description of a capability gap or a technology program) into the individual words that make it up. Second, we stemmed words by removing endings such as *ed* or *s* to retain root words. Third, we removed stop words such as articles (e.g., *the, a, and*) with low semantic meaning. Fourth, we converted records into a "term-document" matrix that contains the number of times each term appears in each record.

Once a term-document matrix is formed, the next step is to populate it with a measure of the occurrence of terms in the documents. We used a common measure called term frequency–inverse document frequency (TF-IDF), which provides a signal of how informative a term is (Ramos, 2003). TF-IDF combines two aspects of frequency: how often a term appears in a particular document (*term frequency*), and how many documents the term appears in (*inverse document frequency*). Intuitively, TF-IDF assigns greater importance to words that appear frequently, but in fewer documents, and it assigns lesser importance to words that appear in many documents. The formula for TF-IDF is

$$TFIDF_{t,d} = f_{t,d} \times log \frac{N}{n_t}$$

where t and d are a given term and document, $f_{t,d}$ is the number of times that the term appears in the document, N is the total number of documents, and n_t is the number of documents containing the term.

Applications of Supervised Learning to Joint Lessons Learned

A basic distinction in ML is between supervised and unsupervised learning. Supervised learning algorithms are trained on input data with labeled outputs, and they are used to predict the classes or values for new test data. Unsupervised learning algorithms are trained on data without labeled outputs; and they are used to group records into clusters, to discover associations between variables, or to reduce the dimensionality of a data set.

Our primary analysis of the unclassified JLLIS data set used NLP algorithms to convert free-text entries into features, and unsupervised learning algorithms (LSA and hierarchical cluster analysis) to group records into semantically related clusters. In this section, we explore the feasibility of using supervised learning techniques to classify whether JLLIS records contain a specific topic of interest. We focus on Air Force Task 5.1, "Provide Airlift Capabilities" (DAF, 1998):

> To organize, train, equip, provide, and plan for the use of forces for air transport for the armed forces. Airlift is the transportation of personnel and materiel through the air and can be applied across the entire range of military operations in support of national objectives. Airlift provides rapid and flexible force mobility options that allow military forces to respond to and operate in a wider variety of circumstances and time frames.

Methodology

Coding

Supervised learning requires labeled data. While JLLIS contains a metadata field to record which Air Force task(s) an entry concerns, the field is rarely filled in. To create a labeled data set, members from the project team manually coded 800 lessons learned according to whether they related to the Air Force task "Provide Airlift Capabilities." Interrater reliability was high.[1]

Preprocessing

Each JLLIS record contains five text fields: topic, observation, discussion, recommendations, and implications. We combined the fields to create a single entry, or "document," for each record. We then applied the sequence of NLP preprocessing steps described above.

Training and Testing

We evaluated two supervised learning algorithms: support vector machines (SVMs) and penalized general linear models (GLMs).[2] In brief, an SVM places each record in a high-dimensional space defined by the feature values (i.e., the TF-IDF values for all word instances). The SVM then discovers a boundary that separates records from different categories. The GLM involves estimating a set of coefficients to convert feature values (i.e., the TF-IDF values for all word instances once again) into the predicted probability that a record belongs to a certain category. In *penalized GLM*, a penalty term is included to shrink regression coefficients for

---

[1] We computed Cohen's kappa, a measure of interrater reliability for qualitative (categorical) items. The value, 0.77, indicated "substantial" interrater reliability. See Landis and Koch, 1977.

[2] We used the e1071 package to implement SVMs, and the glmnet package to implement penalized GLMs.

features with low predictive power to zero. Both approaches are commonly used for prediction tasks including text classification.

Each entry in our data set contained a set of input values from the corresponding row of the TF-IDF matrix, as well as an output label corresponding to whether the entry was related to the Air Force task "Provide Airlift Capabilities." We trained and evaluated predictive models using leave-one-out cross-validation. This involved setting aside one record (i.e., the test set), calculating the TF-IDF values and fitting statistical models using the remaining 799 records (i.e., the training set), and then using the TF-IDF and fitted models to predict the category for the record that was set aside. We repeated this procedure for all 800 records.[3]

For comparison, we also evaluated one unsupervised learning algorithm: LSA. We extracted a set of latent topics from the set of documents and represented each document mathematically as a mixture of the latent topics. We then defined a set of key words to describe the Air Force task "Provide Airlift Capabilities." The key words were *airlift*, *mobility*, *transport*, and *cargo*. We projected the list of key words into the same semantic space as the 800 documents and computed their semantic similarity to each document.

We evaluated predictions using three measures: precision, recall, and F1 score. Precision is the model's accuracy when it predicts that a record relates to airlift. This captures the tendency of a model to selectively flag records that correspond to the topic of interest. Recall is the model's accuracy for true airlift records. This captures the tendency of a model to not miss records that correspond to the topic of interest. Finally, the F1 score is the harmonic mean of precision and recall. This captures the tendency of a model to selectively flag records of interest and to ignore all others.

The GLM and SVM models returned the probability that each record related to airlift. We determined probability threshold values to convert these probabilities into category labels.[4] Additionally, the LSA model returned the semantic similarity between each record and the key word list. Once again, we determined a threshold to convert these scores into category labels. Converting continuous model outputs into binary classifications allowed us to apply standard measures of model performance. However, the continuous values convey additional information about the certainty of classification; records with values closest to the threshold are most ambiguous in terms of their relevance.

---

[3] GLM and SVM both contain metaparameters that control the learning process. We used cross-validation to determine the optimal values for the metaparameters. For the GLM, we set metaparameters to alpha = 1 and lambda = .03. For the SVM, we used a linear kernel and we set cost = .001.

[4] We set thresholds to the values that maximized the F1 score for the GLM and SVM models.

Results

About 7 percent of records in the data set relate to airlift. Table C.1 reports on the performance of the two methods according to the three metrics. Higher values for the metrics indicate better model performance. All three metrics had higher values for GLM than for SVM. Of the records that GLM classified as airlift, 52 percent were correct. In addition, of the records that relate to airlift, GLM correctly classified 76 percent.

To contextualize these results, the expected number of records related to airlift per 1,000 entries equals 68. On average, the GLM model will correctly identify 52 of these entries, and it will miss the remaining 16. In addition, the GLM model will spuriously return 48 records that do not relate to airlift. By adopting a more liberal threshold, the GLM model would miss fewer relevant records, but it would also return more spurious results.

Table C.1. Model Classification Results

Algorithm	Precision	Recall	F1 Measure
GLM	0.52	0.76	0.61
LSA	0.68	0.48	0.57
SVM	0.12	0.69	0.20

To identify the terms that contribute most to model predictions, we computed the correlations between model outputs and the TF-IDF values for each term. Table C.2 displays the top ten terms for the GLM and SVM models. Although the model outputs are based on thousands of possible terms, the fact that these ten are strongly correlated with outputs suggests that they have high importance in the underlying models. Many of the same terms were important in both models, and a human would also rate many of these terms as relating to airlift (e.g., *airlift, cargo, amd, pallet*).

Table C.2. Top Terms That Correlate with Predicted Probabilities in the Support Vector Machine and General Linear Models

GLM	SVM
airlift	cargo
cargo	airlift
amd	arriv
pallet	aircraft
equip	delai
delai	dover
arriv	pallet
turbul	mission
load	rout

NOTE: This list shows high-frequency tokens. Some are words, some are word stems, and some are abbreviations.

Discussion

Given a labeled set of examples, supervised learning techniques can be combined with NLP to return records that are related to a given topic. Once trained, the prediction models can be applied to the complete set of records (e.g., the thousands of warfighter entries contained in JLLIS) and to new records in the future. This approach may be more accurate than unsupervised learning because the human user directly identifies records that contain the collection of words and themes that are most relevant. However, this approach is also more labor intensive because it involves supplying category labels in advance.

The Bidirectional Encoder Representations from Transformers

The language model contained in SCAnTExT uses a "bag-of-words" approach. Each document is broken down into individual words, and information about sentences and surrounding context is discarded. Thus, the simple model does not consider how a word is used, and it does not contain information about the English language more generally. In this section, we describe how a more sophisticated language model could be incorporated into a future version of the tool.

Researchers have recently shown that pretrained language models are effective for many NLP tasks. These models take advantage of the concept of transfer learning. A neural network is extensively trained using a large corpus of generic data and is then fine-tuned for a particular application using a smaller corpus of more specific data. The Bidirectional Encoder Representations from Transformers (BERT) algorithm (Devlin et al., 2019) is one of the most successful pretrained models (Ajayi, 2020).

BERT belongs to the class of masked language models that are trained by randomly masking or hiding words/tokens and having the algorithm predict the masked words based on other unmasked words present in the sentence. As compared with other language models, which parse sentences in a sequential manner, BERT uses a transformer architecture to look simultaneously at all the words in a sentence. This allows the model to learn the context of a word by weighing a contribution from all other words in the sentence (a concept known in NLP as self-attention) (Vaswani et al., 2017).

As a first step toward including more sophisticated language models into SCAnTExT, we demonstrate how BERT can be used to identify technology fields/topics represented in the arXiv NSF, and SBIR databases.

Topic Modeling Using Bidirectional Encoder Representations from Transformers

The goal of the topic modeling is to decompose documents into mixtures of topics. Clustering algorithms can then be used to group documents that address similar topics close to one another.

This is one way to organize and extract meaning from text data. We used BERT to perform topic modeling. Our approach consisted of four steps:

1. Embeddings generation using BERT: Embedding is a numerical representation of text from a document. In BERT, an embedding is a vector with 768 elements that represent the words and context contained in a document. In our application, we generated embeddings for abstracts taken from academic papers submitted to arXiv, NSF awards, and SBIR awards.
2. Dimensionality reduction: After generating embeddings, we reduced the dimensionality of the data set from 768 to 100 dimensions by using Uniform Manifold Approximation and Projection (McInnes, Healy, and Melville, 2020), a nonlinear dimension reduction algorithm that projects high-dimensional structure onto a lower-dimensional space. This step improves the performance of clustering algorithms, which we subsequently applied.
3. Clustering: We used Hierarchical Density-Based Spatial Clustering of Applications with Noise (HDBSCAN; Campello, Moulavi, and Sander, 2013) to discover groups of related documents. HDBSCAN can detect clusters with arbitrary shape and density, it is robust against noisy data, and it does not require an a priori number of clusters.
4. Topic determination: Because the clustering is done using BERT embeddings, the results are not human-interpretable. To overcome this limitation, we used class-based TF-IDF to determine the relevant words that occur more frequently within each cluster. Like TF-IDF, class-based TF-IDF measures the frequency of a word within a cluster relative to the frequency of the word across all clusters.

We applied these steps to a random sample of 60,000 entries from three data sets (i.e., 20,000 entries from the arXiv, NSF, and SBIR databases). The upper panel of Figure C.1 shows a two-dimensional representation of the distribution of documents in BERT-embedded space. NSF and SBIR records (blue and red points) overlap considerably with one another, whereas arXiv records are largely separate. This is consistent with the fact that arXiv papers primarily involve basic scientific research from a narrow range of academic disciplines (e.g., biology, physics, and astronomy), whereas NSF and SBIR awards involve more advanced research from a wider range of academic disciplines.

We applied HDBSCAN to the 60,000 embeddings from the three sources and discovered 389 separate topics (see the lower panel of Figure C.1). We then used class-based TF-IDF to identify words that differed between the clusters. Table C.3 shows the top ten most frequent words using the TF-IDF algorithm for the five largest clusters, which contained 2,522, 1,544, 1,087, 1,027, and 729 documents, respectively. Based on the distinguishing words, clusters 1 through 5 appear to relate to mathematics; cancer treatment; education and science, technology, engineering, and mathematics (STEM); space technology; and quantum physics, respectively. Of these five clusters, the fourth, which relates to space technology, may be most relevant to the TCO in terms of addressing space capability gaps.

Figure C.1. Bidirectional Encoder Representations from Transformers

Embeddings by Source

Embeddings by Cluster

Table C.3. Top Five Topics Among All Databases

Cluster 1	Cluster 2	Cluster 3	Cluster 4	Cluster 5
Math	Patients	Teachers	NASA	Quantum
Prove	Treatment	STEM	NBSP	Spin
Algebra	Clinical	Faculty	Spacecraft	Atoms
Algebras	Patient	Teaching	Missions	Atomic
Mathcal	Care	Student	Space	Magnetic
Let	Therapy	Teacher	Propulsion	Physics
Lie	Cancer	Education	Launch	Electronic
Theorem	Disease	Skills	Cubesat	Electron
Spaces	Pain	Professional	Satellite	Electrons
Finite	Therapeutic	College	Propellant	Topological

NOTE: This table shows some of the most common topics. Some are shown as words, some as word stems, and some as abbreviations.

Discussion

There has been a recent surge of interest in using pretrained language models such as BERT to perform NLP tasks. We explored the use of BERT to identify topic content in different technology data bases. We found that the method returned homogeneous groups of semantically related records. The TCO could use this method to identify clusters of records related to the same technologies, to trace those clusters across sectors to identify technology leaders, and to trace those clusters across time to estimate technology maturity. In addition, by placing records related to capability gaps and technology solutions in a common semantic space, the method could identify technologies that are semantically related and potentially relevant to a particular capability gap.

Appendix D. Semantic Clustering Analysis and the Thematic Exploration Tool

SCAnTExT is implemented in the R computing language and can be run locally or hosted on a Shiny Server. SCAnTExT may also be run on classified computing systems with R installed. The primary cost for maintaining the tool comes from the time to gather and prepare new records and data sources as they become available.

Data Sources

SCAnTExT contains 15 unclassified and classified data sources (see Table D.1). The same NLP path is applied to all sources. Sources differ in terms of whether they primarily describe operational and technology capability gaps, technology solutions, or both. Sources also differ in terms of the metadata they contain. For example, besides titles and descriptions, records from different data sources may include funding amount, funding source, technological readiness level, company, and point of contact, among other things.

More records are available from the data sources that we compiled. For example, JLLIS contains Army and Navy lessons learned; and arXiv, DTIC, NSF, RDT&E, SBIR, and STTR records are available from earlier years. These records can be added to SCAnTExT by applying the existing cleansing scripts and NLP path. New data sources can also be added. The initial steps of retrieving, digitizing, and parsing a new source and identifying words and phrases to exclude from its entries are unique for each source. Once these steps are complete, the NLP path can be applied to new data sources, and the sources can be added to SCAnTExT without modifying the tool.

Table D.1. Data Sources Contained in SCAnTExT

Type	Name	Records	Metadata Fields	Notes
Capability gaps	USAF JLLIS (unclassified)	6,929	Service; organization; date; point of contact (POC)	2001–present; available for other services
	USAF JLLIS (classified)	2,503	Service; organization; date; POC	2001–present; available for other services
	AFUTL (unclassified)	280		
	CBA for health services support to agile combat employment CONOPS in PACAF (unclassified)	81	Gap characterization; category; subcategory; task	
	Air Mobility Command, Office of the Surgeon General En Route Care CBA Report (unclassified)	99	Core capability area; gap characterization	
	CCMD Common Capability Needs, February 2020 (unclassified)	118	Command; category; subcategory	
	CGLs and IPLs (classified)	268		9 MAJCOMs and CCMDs
Technology solutions	AFRL FY 2022 POM (unclassified)	619	Budget activity; lead organization; primary core technical competency; amount; POC; FY	
	arXiv abstracts (unclassified)	40,000	Category; date; POC	Sample from 2020; additional records and earlier years available
	AFVentures portfolio	1,358	Technology category	Company descriptions and key words
	FY 2022 research, development, test, and evaluation defense budget materials (unclassified)	2,673	Budget activity; office; amount; FY	Program elements and projects
	NSF proposals	32,256	Directorate; division; amount; year; POC	2015–present; earlier years available
	SBIR and STTR awards (unclassified)	31,045	Firm; funding agency; phase; program; amount; FY; POC	2016–present; earlier years available
	DTIC abstracts	31,632	Category type; TRL; community of interest; Amount; POC; FY	2016–present; earlier years available
Both	Federally funded research and development reports	256	Program; FY	

User Interface

Figure D.1 shows the user interface that appears when the web application is launched. The SCAnTExT overview describes the tool and its functionality.

Figure D.1. SCAnTExT Overview

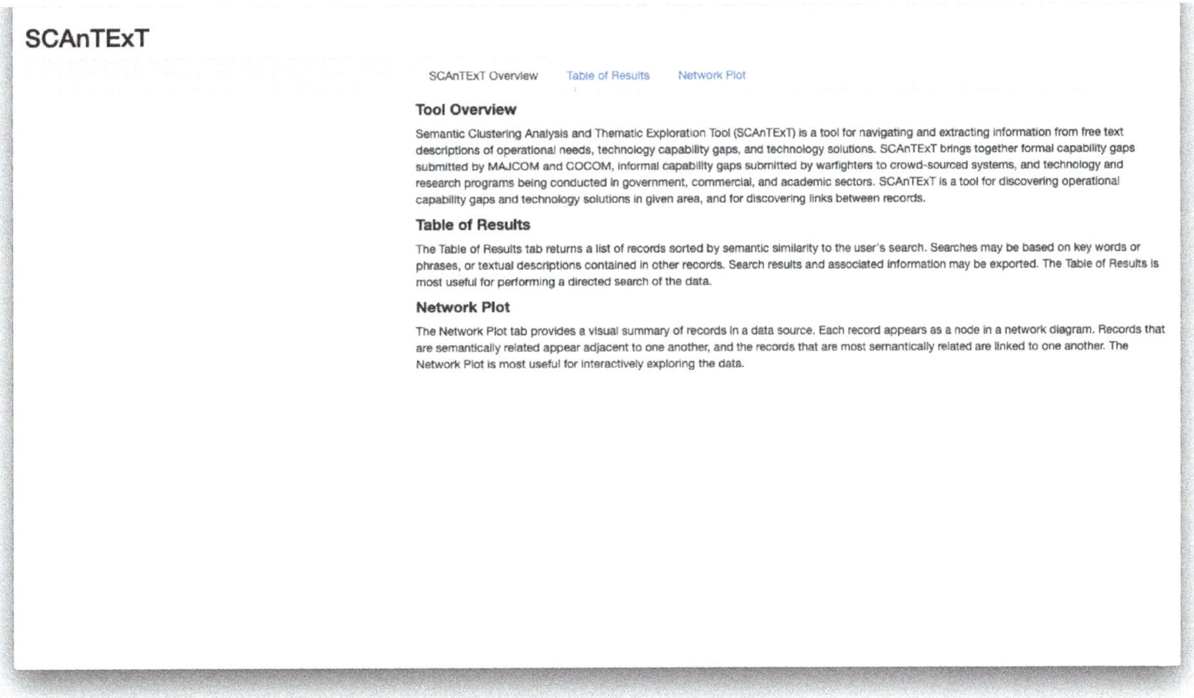

Figure D.2 shows the "table of results" tab. This tab allows the user to search records in multiple ways. A list of records sorted by semantic similarity to the user's search are returned in the table. The search results may be exported for offline processing.

The "table of results" tab contains multiple elements and accepts multiple user inputs:

- **Which data set do you want to search through?** The user may select one of the data sets (see Table D.1) to search through. Searches return records from this data set.
- **Which data set do you want to seed your search from?** The user may designate one of the data sets (see Table D.1) as the reference data set. The user may select records or groups of records from this data set and use them to find related records in the searchable data set.
- **Select a record that you want to match (optional).** The user may select one or more records from the reference data set to return semantically related records from the searchable data set.

85

- **Select a group of records that you want to match (optional).** The user may select clusters of records from the reference data set to return semantically related records from the searchable data set.
- **Provide a search phrase (optional).** The user may enter key words and phrases to return semantically related records from the searchable data set.
- **Download table.** The user may select rows in the table and download a file that contains record titles, descriptions, and other information.

The example in Figure D.2 shows how a user can retrieve results related to a specific topic. In the example, the user sets RDT&E budget documentation as the searchable data set. The records in the table are from DoD-wide PEs and projects described in the RDT&E budget documentation. The user types the search terms *cyber operations*. The records in the table are sorted by their semantic similarity to the search terms. The user may hover over a row in the table to read the record's complete description, and the user may select one or more rows to export.

Figure D.2. SCAnTExT Table of Results Tab Search by Text

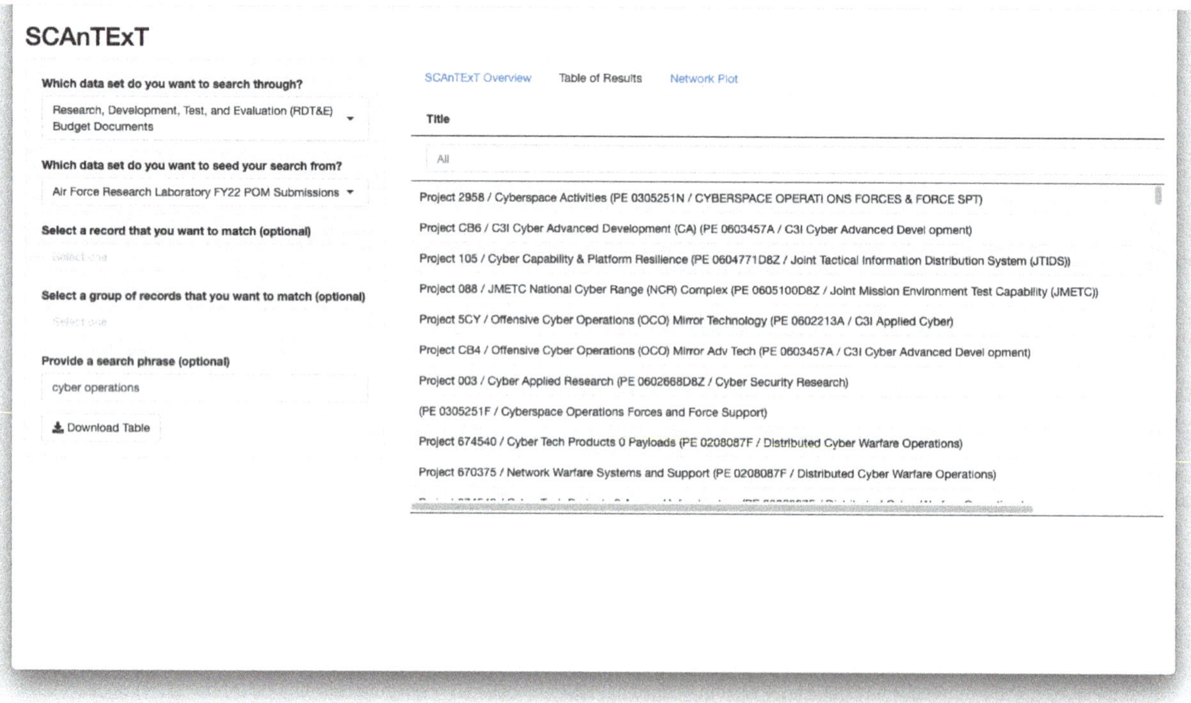

The example in Figure D.3 shows how a user can retrieve results related to another record. In the example, the user sets the reference data set to a collection of AFRL program descriptions, and the searchable data set to RDT&E budget documentation. In the field, "Select a record that you want to match (optional)," the user selects an AFRL research program titled "ACC2: Adaptive Cyber Command and Control." The records in the table are RDT&E PEs and projects sorted by their semantic similarity to the description of the AFRL ACC2 program. Once again, the user can access and export additional information about the records in the table.

Figure D.3. SCAnTExT Table of Results Tab Search by Reference Data Set Record

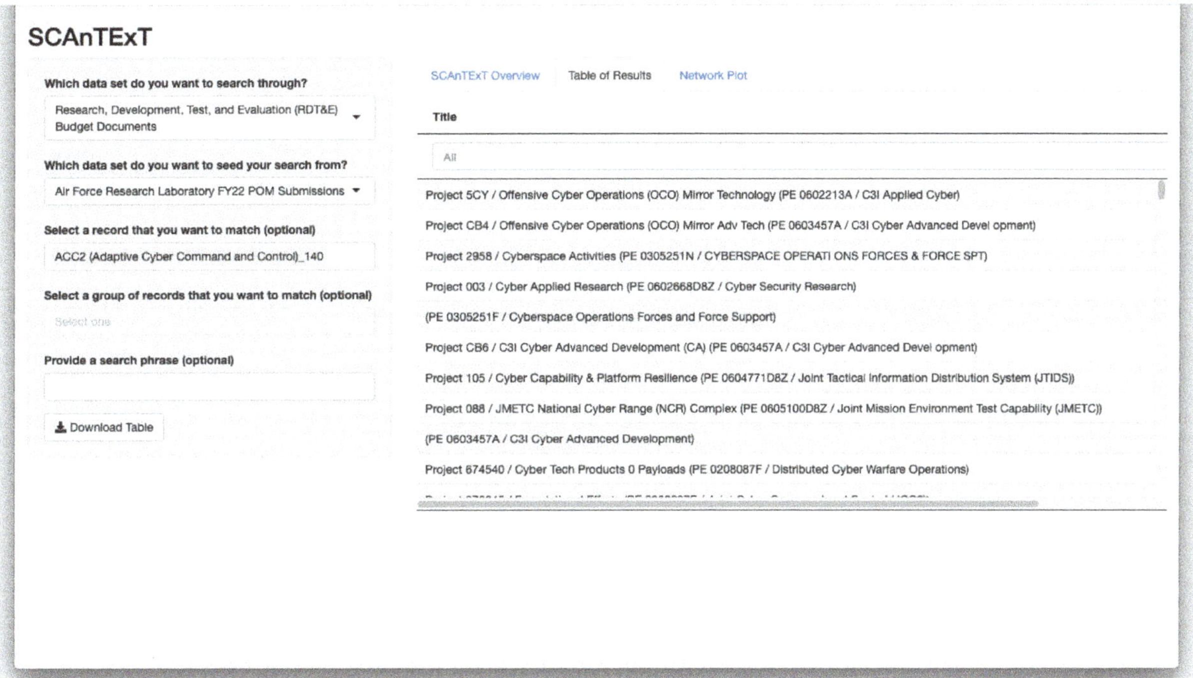

Figure D.4 shows the "network plot" tab. This tab allows the user to visualize the relationships between records in a data set as a network diagram. Each node in the diagram is one of the 2,673 PE and project entries from the RDT&E documentation. Records that are semantically related appear in the same color and are near one another, and the subset of records that are most semantically similar are linked.

Figure D.4. SCAnTExT Network Plot Tab

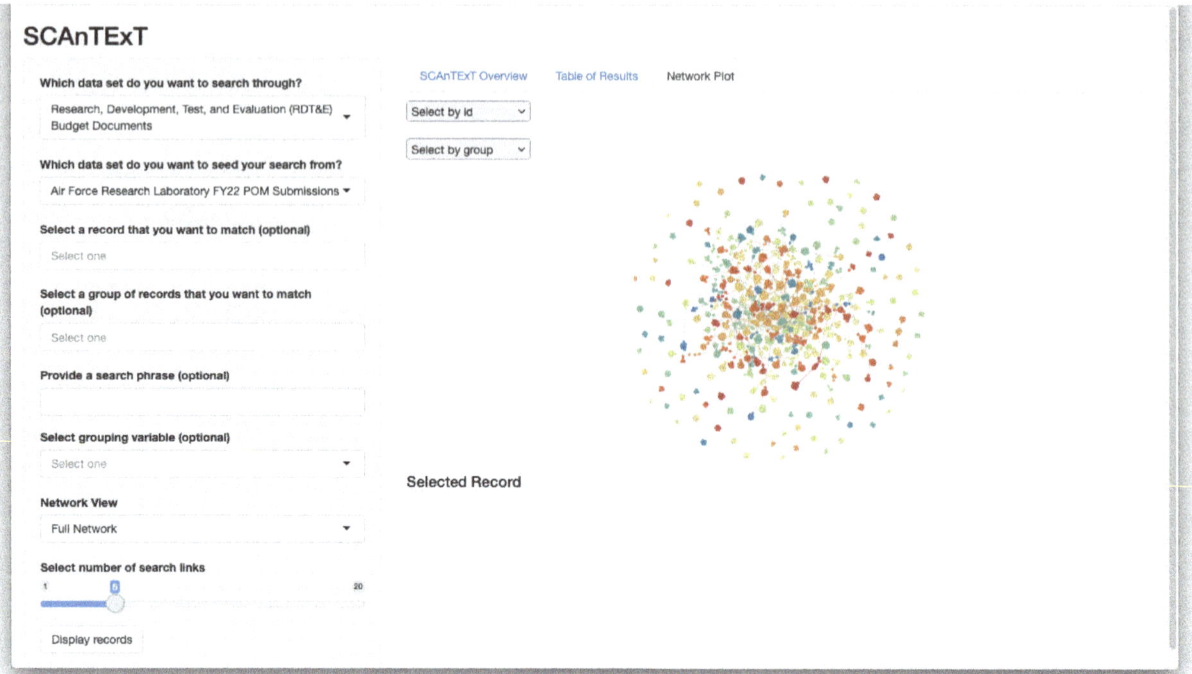

The "network plot" tab contains multiple elements and accepts multiple user inputs. Some of the elements also appear in the "table of results" tab, and some are new:

- **Select grouping variable (optional).** By default, the color of a node in the network diagram is determined by the semantic cluster that it belongs to. The user may select other meta tags like *military service*, *lead organization*, *budget activity*, or *phase* to set node color.
- **Network view.** By default, the complete network diagram is shown. For larger data sets, the user may select to only view the set of nodes most closely related to the search inputs.
- **Select number of search links.** When search inputs are provided, a search node appears outside the network diagram. The number of search links determines how many nodes in the network are linked to the search node (i.e., the *n* most semantically similar records).
- **Display records.** The user may select and display all information associated with a node.

Figure D.5 gives an example of how a user can search the network diagram. In the example, the user sets RDT&E budget documentation to the searchable data set and types the search term *cyber operations*. The blue node appears outside the network and is connected to the 20 most semantically similar records in the network. These are the same PEs and projects listed at the top of the table in Figure D.2. The user could select any of the nodes in the network diagram to find additional records with high semantic similarity. Finally, by changing the grouping variable to service, the user could see the number of cyber-related programs sponsored by different DoD offices.

Figure D.5. SCAnTExT Network Plot Tab Search by Text

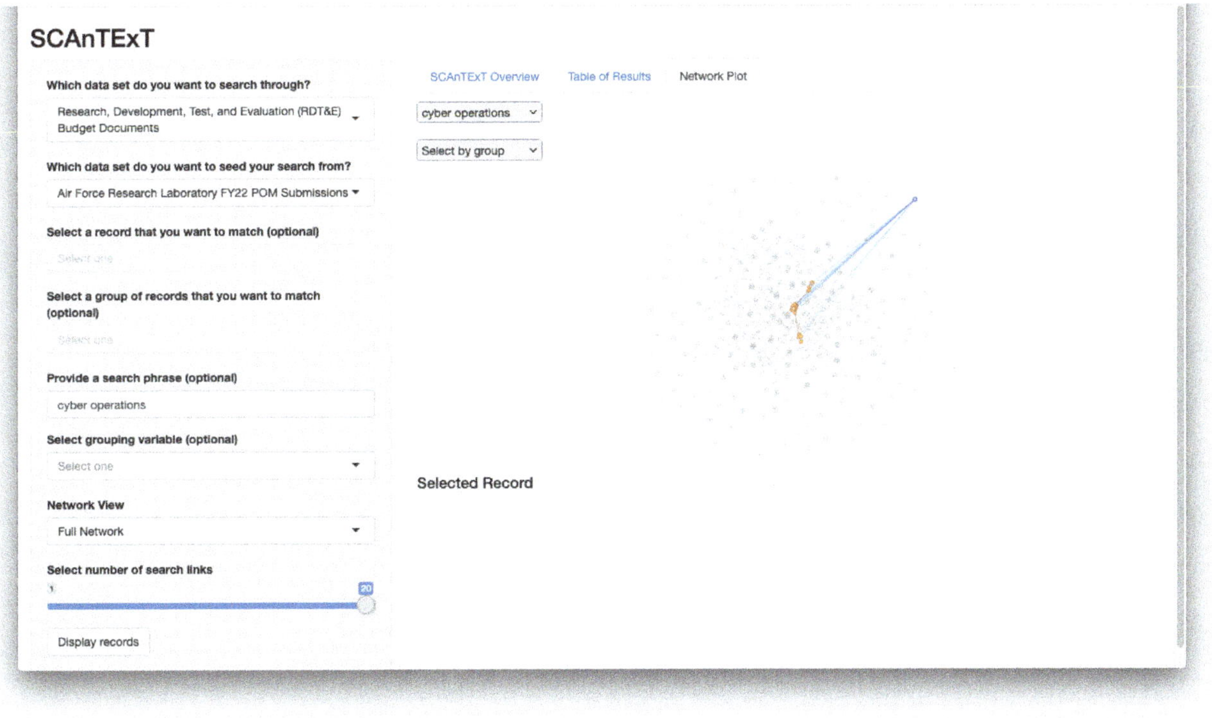

SOURCE: SCAnTExT web application.

Abbreviations

ACC	Air Combat Command
ACC2	Adaptive Cyber Command and Control
AE	aeromedical evacuation
AFB	Air Force Base
AFMLO	Air Force Medical Logistics Office
AFRL	Air Force Research Laboratory
AF-SMP	Air Force Strategic Master Plan
AFSOC	Air Force Special Operations Command
AFUTL	Air Force Universal Task List
AFWIC	Air Force Warfighter Integration Capability
AI	artificial intelligence
AOC	Air Operations Center
API	Airmen Powered by Innovation
AQR	Air Force Acquisition: Science, Technology, and Engineering Directorate
BERT	Bidirectional Encoder Representations from Transformers
C2	command and control
CASF	Contingency Aeromedical Staging Facility
CBA	Capabilities Based Assessment
CCATT	Critical Care Air Transport Team
CCMD	Combatant Command
CGL	capability gap list
CONOPS	concept of operations
CSAR	combat search and rescue
DAF	Department of the Air Force
DoD	Department of Defense
DTIC	Defense Technical Information Center
EMDG	Expeditionary Medical Group

EUCOM	European Command
FY	fiscal year
GLM	general linear model
GPS	global positioning system
HCDE	human-centered, data-enhanced
HCM	human capital management
HDBSCAN	Hierarchical Density-Based Spatial Clustering of Applications with Noise
HSVTOL	high-speed vertical takeoff and landing
IPL	integrated priority list
IT	information technology
JADC2	Joint All-Domain Command and Control
JLLIS	Joint Lessons Learned Information System
LIMFAC	limiting factor
LSA	latent semantic analysis
MAJCOM	Major Command
MASF	Mobile Aeromedical Staging Facility
ML	machine learning
NASA	National Aeronautics and Space Administration
NDS	National Defense Strategy
NLP	natural language processing
NLU	natural language understanding
NMS	National Military Strategy
NSF	National Science Foundation
PACAF	Pacific Air Forces
PAF	RAND Project AIR FORCE
PE	program element
PEO	program executive officer
PMR	Patient Movement Request
POC	point of contact
POM	Program Objective Memorandum

PR	personnel recovery
R&D	research and development
RDT&E	research, development, test, and evaluation
S&T	science and technology
SBIR	Small Business Innovation Research
SCAnTExT	Semantic Clustering Analysis and Thematic Exploration Tool
SME	subject matter expert
SMIC	Systèmes et Matrices d'Impacts Croisés
STEM	science, technology, engineering, and mathematics
STIPL	science and technology integrated priority list
STTR	Small Business Technology Transfer
SVM	support vector machine
TCO	Transformational Capabilities Office
TEO	technology executive officer
TF-IDF	term frequency–inverse document frequency
TRL	technology readiness level
TSA	time series analysis
USAF	U.S. Air Force

References

AFRL—*See* Air Force Research Laboratory.

AFSOC—*See* Air Force Special Operations Command.

AFWERX, HSVTOL Concept Challenge, webpage, undated. As of September 14, 2021:
https://afwerxchallenge.com/air/highspeedvtol/overview

Air Force Futures, *Our Goals for 2021*. Washington, D.C., 2021.

Air Force Research Laboratory, "Air Force Vanguards," website, undated. As of September 21, 2021:
https://afresearchlab.com/technology/vanguards/

Air Force Special Operations Command, *Concept of Employment: Air Force Special Operations Command. High Speed Vertical Takeoff and Landing (HSVTOL) Aircraft*, Washington, D.C., 2021.

Ajayi, Demi, "How BERT and GPT Models Changed the Game for NLP," *IBM Watson Blog*, December 3, 2020. As of September 30, 2021:
https://www.ibm.com/blogs/watson/2020/12/how-bert-and-gpt-models-change-the-game-for-nlp/

Attride-Stirling, Jennifer, "Thematic Networks: An Analytic Tool for Qualitative Research," *Qualitative Research*, Vol. 1, No. 3, December 2001, pp. 385–405. As of October 1, 2021:
https://journals.sagepub.com/doi/10.1177/146879410100100307

Balakrishnan, Tara, Michael Chui, Bryce Hall, and Nicolaus Henke, *The State of AI in 2020*, McKinsey & Company, November 17, 2020. As of October 1, 2021:
https://www.mckinsey.com/business-functions/mckinsey-analytics/our-insights/global-survey-the-state-of-ai-in-2020

Boe-Lillegraven, Siri, and Stephan Monterde, "Exploring the Cognitive Value of Technology Foresight: The Case of the Cisco Technology Radar," *Technological Forecasting and Social Change*, Vol. 101, December 2015, pp. 62–82. As of September 30, 2021:
https://www.sciencedirect.com/science/article/pii/S004016251400239X

Braun, Virginia, and Victoria Clarke, "Using Thematic Analysis in Psychology," *Qualitative Research in Psychology*, Vol. 3, No. 2, 2006, pp. 77–101. As of October 1, 2021:
https://www.tandfonline.com/doi/abs/10.1191/1478088706qp063oa

Brem, Alexander, and Kai-Ingo Voigt, "Integration of Market Pull and Technology Push in the Corporate Front End and Innovation Management—Insights from the German Software Industry," *Technovation*, Vol. 29, No. 5, May 2009, pp. 351–367. As of October 1, 2021: https://www.researchgate.net/publication/239358064_Integration_of_Market_Pull_and _Technology_Push_in_the_Corporate_Front_End_and_Innovation_Management_ -_Insights_from_the_German_Software_Industry

Brown, Charles Q. Jr., *CSAF Action Orders to Accelerate Change Across the Air Force*, Air Force, webpage, December 10, 2020. As of November 21, 2022: https://www.af.mil/Portals/1/documents/csaf/CSAF_Action_Orders_Letter_to_the_Force.pdf

Cambria, Erik, and Bebo White, "Jumping NLP Curves: A Review of Natural Language Processing Research," *IEEE Computational Intelligence Magazine*, Vol. 9, No. 2, May 2014, pp. 48–57. As of October 1, 2021: https://www.semanticscholar.org/paper/Jumping-NLP-Curves%3A-A-Review-of-Natural -Language-Cambria-White/ec43d37aad744150af144d27a08b0b097607e712

Campello, Ricardo J. G. B., Davoud Moulavi, and Joerg Sander, "Density-Based Clustering Based on Hierarchical Density Estimates," in Jian Pei, Vincent S. Tseng, Longbing Cao, Hiroshi Motoda, and Guandong Xu, eds., *Advances in Knowledge Discovery and Data Mining: 17th Pac-Asia Conference, PAKDD 2013, Gold Coast, Australia, April 2013*, Part II, Berlin: Springer, 2013, pp. 160–172.

Christensen, Clayton M., Michael E. Raynor, and Rory McDonald, "What Is Disruptive Innovation?" *Harvard Business Review*, Vol. 93, No. 12, December 2015, pp. 44–53. As of October 1, 2021: https://hbr.org/2015/12/what-is-disruptive-innovation

Cornell University, "arXiv Monthly Submissions," web chart, last updated October 1, 2021. As of October 1, 2021: https://arxiv.org/stats/monthly_submissions

Cozzens, Susan, Sonia Gatchair, Jongseok Kang, Kyung-Sup Kim, Hyuck Jai Lee, Gonzalo Ordóñez, and Alan Porter, "Emerging Technologies: Quantitative Identification and Measurement," *Technology Analysis & Strategic Management*, Vol. 22, No. 3, 2010, pp. 361–376. As of October 1, 2021: https://www.tandfonline.com/doi/full/10.1080/09537321003647396

DAF—*See* Department of the Air Force.

Department of the Air Force, *Air Force Task List (AFTL)*, Washington, D.C.: Air Force Doctrine Document 1-1, August 12, 1998. As of September 30, 2021: https://apps.dtic.mil/sti/citations/ADA356428

———, *USAF Strategic Master Plan*, Washington, D.C., May 2015. As of September 21, 2021:
https://www.af.mil/Portals/1/documents/Force%20Management/Strategic_Master_Plan.pdf

———, *Science and Technology Strategy: Strengthening USAF Science and Technology for 2030 and Beyond*, Washington, D.C., April 2019. As of September 21, 2021:
https://www.af.mil/Portals/1/documents/2019%20SAF%20story%20attachments/Air%20Force%20Science%20and%20Technology%20Strategy.pdf

Devlin, Jacob, Ming-Wei Chang, Kenton Lee, and Kristina Toutanova, *BERT: Pre-Training of Deep Bidirectional Transformers for Language Understanding*, arXiv:1810.04805v2 [cs.CL], May 24, 2019. As of September 30, 2021:
https://arxiv.org/abs/1810.04805

DoD—*See* U.S. Department of Defense.

Greenblott, Joseph M., Thomas O'Farrell, Robert Olson, and Beth Burchard, "Strategic Foresight in the Federal Government: A Survey of Methods, Resources, and Institutional Arrangements," *World Futures Review*, Vol. 11, No. 3, 2019, pp. 245–266. As of September 30, 2021:
https://journals.sagepub.com/doi/full/10.1177/1946756718814908

Gulbrandsen, Karen Elizabeth, *Bridging the Valley of Death: The Rhetoric of Technology Transfer*, dissertation, Iowa State University, 2009. As of September 30, 2021:
https://lib.dr.iastate.edu/etd/10740

Hammoud, Mohamad Saleh, and Douglas P. Nash, "What Corporations Do with Foresight," *European Journal of Futures Research*, Vol. 2, No. 1, June 21, 2014, pp. 1–20. As of September 30, 2021:
https://eujournalfuturesresearch.springeropen.com/articles/10.1007/s40309-014-0042-9

Joint Chiefs of Staff, "Joint Lessons Learned Information System," webpage, undated. As of October 1, 2021:
https://www.jcs.mil/Doctrine/Joint-Lessons-Learned/

Jordan, Michael I., and Tom M. Mitchell, "Machine Learning: Trends, Perspectives, and Prospects," *Science*, Vol. 349, No. 6245, July 17, 2015, pp. 255–260. As of October 1, 2021:
https://www.science.org/lookup/doi/10.1126/science.aaa8415

Keenan, Michael, Rémi Barré, and Cristiano Cagnin, "Future-Oriented Technology Analysis: Future Directions," in Cristiano Cagnin, Michael Keenan, Ron Johnston, Fabiana Scapolo, and Rémi Barré, eds., *Future-Oriented Technology Analysis: Strategic Intelligence for an Innovative Economy*, Berlin: Springer, 2008, pp. 163–169. As of September 13, 2022:
https://link.springer.com/chapter/10.1007/978-3-540-68811-2_12

Kline, Stephen J., and Nathan Rosenberg, "An Overview of Innovation," in *Studies on Science and the Innovation Process: Selected Works of Nathan Rosenberg*, Singapore: World Scientific Publishing, 2010, pp. 173–203. As of September 13, 2022: https://www.worldscientific.com/doi/10.1142/9789814273596_0009

Knopman, Debra S., Don Snyder, Irv Blickstein, David E. Thaler, James Leftwich, Colby P. Steiner, Quentin E. Hodgson, Elaine Simmons, Krista Romita Grocholski, and Yvonne K. Crane, *Proposed Analytical Products for the Air Force Warfighting Integration Capability: Developing and Presenting Options for Future Force Design and Capability Development*, Santa Monica, Calif.: RAND Corporation, RR-4199-AF, 2020. As of October 1, 2021: https://www.rand.org/pubs/research_reports/RR4199.html

Landauer, Thomas K., Peter W. Foltz, and Darrell Laham, "An Introduction to Latent Semantic Analysis," *Discourse Processes*, Vol. 25, No. 23, 1998, pp. 259–284.

Landis, J. Richard, and Gary G. Koch, "The Measurement of Observer Agreement for Categorical Data," *Biometrics*, Vol. 33, No. 1, March 1977, pp. 159–174.

Leftwich, James A., Debra Knopman, Jordan R. Fischback, Michael J. Vermeer, Kristin Van Abel, and Nidhi Kalra, *Air Force Capability Development Planning: Analytical Methods to Support Investment Decisions*, Santa Monica, Calif.: RAND Corporation, RR-2931, 2019. As of September 30, 2021: https://www.rand.org/pubs/research_reports/RR2931.html

Lofgren, Eric, "Analysis: What Does the New AFWERX Data Say About the Tech Pipeline?" *Defense News*, September 11, 2020. As of October 1, 2021: https://www.defensenews.com/opinion/commentary/2020/09/11/analysis-what-does-the-new-afwerx-data-say-about-the-tech-pipeline/

McInnes, Leland, John Healy, and James Melville, *UMAP: Uniform Manifold Approximation and Projection for Dimension Reduction*, arXiv:1802.03426v3, September 18, 2020. As of September 30, 2021: https://arxiv.org/abs/1802.03426

Metz, Steven K., and James O. Kievit, *Strategy and the Revolution in Military Affairs: From Theory to Policy*, Carlisle Barracks, PA: Army War College Strategic Studies Institute, June 27, 1995. As of September 30, 2021: https://apps.dtic.mil/sti/citations/ADA298198

National Research Council, *Persistent Forecasting of Disruptive Technologies*, Washington, D.C.: National Academies Press, 2010. As of October 1, 2021: https://www.nap.edu/catalog/12557/persistent-forecasting-of-disruptive-technologies

———, *Development Planning: A Strategic Approach to Future Airforce Capabilities*, Washington, D.C.: National Academies Press, 2014. As of September 21, 2021: https://www.nap.edu/catalog/18971/development-planning-a-strategic-approach-to-future -air-force-capabilities

Popper, Rafael, "Foresight Methodology," in Luke Georghiou, Jennifer C. Harper, Michael Keenan, Ian Miles, and Rafael Popper, eds., *The Handbook of Technology Foresight*, Northampton, Mass.: Edward Elgar, 2008, pp. 44–88.

Porter, Alan L., "Technology Foresight: Types and Methods," *International Journal of Foresight and Innovation Policy*, Vol. 6, Nos. 1–3, 2010, pp. 36–45.

Pringle, Heather L., "Implementation of the 2030 Science and Technology Strategy Across the Department of the Air Force," presentation to the Armed Services Committee of the United States Senate, Subcommittee on Emerging Threats and Capabilities, April 21, 2021. As of September 14, 2021: https://www.armed-services.senate.gov/imo/media/doc/Pringle%20Testimony%2004 .21.21.pdf

Public Law 115-232, John S. McCain National Defense Authorization Act for Fiscal Year 2019, August 13, 2018. As of October 1, 2021: https://www.govinfo.gov/app/details/BILLS-115hr5515enr/summary

Ramos, Juan, "Using TF-IDF to Determine Word Relevance in Document Queries," in *Proceedings of the First Instructional Conference on Machine Learning*, Piscataway, NJ: Association for Computing Machinery, 2003, pp. 133–142. As of November 21, 2022: https://www.semanticscholar.org/paper/Using-TF-IDF-to-Determine-Word-Relevance-in -Queries-Ramos/b3bf6373ff41a115197cb5b30e57830c16130c2c

R Foundation. "The R Project for Statistical Computing," homepage, undated. As of September 14, 2022: https://www.R-project.org/

Rohrbeck, Rene, and Hans Georg Gemünden, "Corporate Foresight: Its Three Roles in Enhancing the Innovation Capacity of a Firm," *Technological Forecasting and Social Change*, Vol. 78, No. 2, January 2010, pp. 231–243. As of September 30, 2021: https://www.researchgate.net/publication/202288905_Corporate_Foresight_Its_Three _Roles_in_Enhancing_the_Innovation_Capacity_of_a_Firm

Russel, Stuart, and Peter Norvig, *Introduction to Artificial Intelligence: A Modern Approach*, Englewood Cliffs, N.J.: Prentice Hall, 1995.

Sargent, John F., Jr., *Department of Defense Research, Development, Test, and Evaluation (RDT&E): Appropriations Structure*, Washington, D.C., Congressional Research Service, R44711, version 9, updated October 7, 2020. As of September 14, 2021:
https://crsreports.congress.gov/product/details?prodcode=R44711

Schatzmann, Jörg, René Schäefer, and Frederick Eichelbaum, "Foresight 2.0: Definition, Overview, and Evaluation," *European Journal of Futures Research*, Vol. 1, No. 1, December 2013, pp. 1–15. As of September 30, 2021:
https://www.researchgate.net/publication/277415924_Foresight_20_-_Definition _overview_evaluation

Schirmer, Pete, Amber Jaycocks, Sean Mann, William Marcellino, Luke J. Matthews, John David Parsons, and David Schulker, *Natural Language Processing: Security- and Defense-Related Lessons Learned*, Santa Monica, Calif.: RAND Corporation, PE-A926-1, 2021. As of September 30, 2021:
https://www.rand.org/pubs/perspectives/PEA926-1.html

Under Secretary of the Air Force and Vice Chief of Staff, October 3, 2017, "Memorandum on the Air Force Warfighting Integration Capability (AFWIC) IOC + Resourcing," Washington, D.C., October 3, 2017.

U.S. Air Force Scientific Advisory Board, *Future Air Force Vanguard Selection and Management Processes: SAB "Quick-Look" Study Outbrief*, Washington, D.C., 2020. As of September 21, 2021:
https://www.scientificadvisoryboard.af.mil/Portals/73/Studies/FY20/SAB%20FY20 %20Vanguard%20Briefing_Distro%20A%20(Final).pdf?ver=tkXwLpZs2DoSmsELblvf3A %3D%3D

U.S. Department of Defense, *Fiscal Year (FY) 2022 Budget Estimates: Air Force; Justification Book*: Vol. 3b, *Research, Development, Test & Evaluation, Air Force*, Washington, D.C.: May 2021.

Vaswani, Ashish, Noam Shazeer, Niki Parmar, Jakob Uszkoreit, Llion Jones, Aidan N. Gomez, Łukasz Kaiser, and Illia Polosukhin, "Attention Is All You Need," in Isabelle Guyon, Ulrike von Luxburg, Samy Bengio, Hanna M. Wallach, Rob Fergus, S. V. N. Vishwanathan, and Roman Garnett, eds., *Advances in Neural Information Processing Systems*: 31st Annual Conference on Neural Information Processing Systems (NIPS 2017), San Diego, Calif.: Neural Information Processing Systems Foundation, pp. 5998–6008. As of September 30, 2021:
https://papers.nips.cc/paper/2017/hash/3f5ee243547dee91fbd053c1c4a845aa-Abstract.html

Vergun, David, "DOD Looking for Advanced Command, Control Solution," U.S. Department of Defense, June 4, 2021. As of September 30, 2021:
https://www.defense.gov/Explore/News/Article/Article/2646822/dod-looking-for-advanced-command-control-solution/

Warwick, Graham, "DARPA Aims to Build Bridge Across Technology's 'Valley of Death,'" Aviation Week Network, August 16, 2021. As of September 30, 2021:
https://aviationweek.com/aerospace/aircraft-propulsion/darpa-aims-build-bridge-across-technologys-valley-death

Wilson, H. James, and Paul R. Daugherty, "The Next Big Breakthrough in AI Will Be Around Language," *Harvard Business Review*, September 23, 2020. As of September 1, 2021:
https://hbr.org/2020/09/the-next-big-breakthrough-in-ai-will-be-around-language

Ye, Chunlei, and Lu Feng, "Future-Oriented Technology Analysis of Technology Roadmap Based on Text Mining," in *Proceedings of the 10th International Conference on Fuzzy Systems and Knowledge Discovery*, FSKD 2013, Shenyang, China, July 23–25, 2013, Piscataway, NJ: Institute of Electrical and Electronics Engineers, 2013, pp. 1126–1130.

Young, Tom, Devamanyu Hazarika, Soujanya Poria, and Erik Cambria, "Recent Trends in Deep Learning Based Natural Language Processing," *IEEE Computational Intelligence Magazine*, Vol. 13, No. 3, August 2018, pp. 55–75. As of September 16, 2021:
https://ieeexplore.ieee.org/document/8416973

Zhang, Daniel, Nestor Maslej, Erik Brynjolfsson, John Etchemendy, Terah Lyons, James Manyika, Helen Ngo, Juan Carlos Niebles, Michael Sellitto, Ellie Sakhaee, Yoav Shoham, Jack Clark, and Raymond Perrault, *Artificial Intelligence Index Report 2022*, AI Index Steering Committee, Stanford, Calif.: Stanford Institute for Human-Centered Artificial Intelligence, March 2022. As of April 4, 2022:
https://aiindex.stanford.edu/report/

Ingram Content Group UK Ltd.
Milton Keynes UK
UKHW050932020623
422767UK00005B/17